# Thermal Flow in Porous Media

Theory and Applications of Transport in Porous Media

*A Series of Books Edited by Jacob Bear, Technion – Israel Institute of Technology, Haifa*

Horia I. Ene and Dan Poliševski

*Dept. of Mathematics, The National Institute for Scientific and Technical Creation, Bucharest, Romania*

# Thermal Flow in Porous Media

D. Reidel Publishing Company

A MEMBER OF THE KLUWER ACADEMIC PUBLISHERS GROUP

Dordrecht / Boston / Lancaster / Tokyo

**Library of Congress Cataloging in Publication Data**

Ene, Horia I.
Thermal flow in porous media.
(Theory and applications of transport in porous media)
Bibliography: p.
Includes index.
1. Fluid dynamics. 2. Heat – Convection,
Natural. 3. Multiphase flow. 4. Porous
materials – Thermal properties. I. Poliševski, Dan. II. Title.
QC155.E54 1987 536'.25 86-31661
ISBN 90-277-2225-0

---

Published by D. Reidel Publishing Company,
P.O. Box 17, 3300 AA Dordrecht, Holland.

Sold and distributed in the U.S.A. and Canada
by Kluwer Academic Publishers
101 Philip Drive, Norwell, MA 02061, U.S.A.

In all other countries, sold and distributed
by Kluwer Academic Publishers Group,
P.O. Box 322, 3300 AH Dordrecht, Holland.



Printed in The Netherlands

# Contents

# Introduction

The transport of heat through a porous medium in the presence of exterior forces, generally produced by the Earth's gravitational field and/or a pressure gradient, is called conduction when the Darcean fluid is static (motionless), and convection when the Darcean fluid is in motion. It is customary to use the term convection also to describe the motion which arises from the density differences due to temperature gradients within the Darcean fluid. We think that because this last phenomenon is more general it should be given a specific name; here we call it thermal flow.

In the sense of the above definitions, convection and thermal flow are two distinct phenomena (they occur together, in underground combustion for instance), and the convective motion which arises when a Darcean fluid is in contact with a source of heat is a particular case of thermal flow.

Thermal flow occurs naturally and is important in many geophysical and industrial problems, particularly in oil exploration, and in the petroleum, chemical and nuclear industries (for instance, in the evaluation of capability of heat-removal from a hypothetical accident in a nuclear reactor). It can play a part in the transfer of heat from the deep interior of the Earth to a shallow depth in the geothermal regions. However, in the field of energy conversion little attention has yet been paid to the insulating characteristics of the saturated porous materials introduced in some enclosures (storage tanks) to decrease the convective and radiative transfer of heat.

This book, then, deals with thermal flow in porous media. Its title has been chosen deliberately in an attempt to establish a new term.

The Darcean fluid does not always consist of a single phase only; accordingly the book is divided into two parts, the first dealing with single-phase flow, and the second with multiphase flow.

The study of single-phase thermal flow through a porous medium has revealed the basic mechanism which governs the phenomenon. For a long time, the Boussinesq assumption—that the density of the Darcean fluid is constant everywhere except in the forcing term (the buoyancy force) which is the cause of the thermal flow—has been adopted and used without any reasonable justification other than the expected similarity with the flow of viscous fluids. However, as we shall show, there are many situations in which this similarity does not hold. The first accurate establishment of the macroscopic equations of the Darcy – Boussinesq system which in a certain approximation governs the thermal flow of viscous fluids through a periodic model of a porous medium have been obtained by using the multiple scale method of homogenization theory. These equations, used together with the convergence theorem of the homogenization process, provide the only means of analysing single-phase thermal flow. The main field of application is natural convection.

The movement of a multiphase Darcean fluid is obviously a much more complex phenomenon. It can be described only by starting from the thermodynamics of the irreversible processes applied to mixtures. But even here phenomena like diffusion and chemical reactions introduce many complications, and a rigorous approach is required. An averaging method is adopted in which a completely heterogeneous medium is replaced by one which conforms to the general principles of mechanics. The system of equations obtained in this way is very complicated, and capable of solution only by using a computer. Typical examples of such problems occur in the analysis of underground combustion, the remarkable oil recovery technique.

In both parts of the book the macroscopic equations obtained rigorously by averaging are used to form a quite general mathematical model of thermal flow. We have therefore considered various particular cases, the geometry of the domains and the corresponding boundary value problems giving the specific aspects of the phenomena. This is not, then, an exhaustive study; in fact it would be an impossible task to include in a book of reasonable size all known results. The present work is intended to be a mathematical foundation of this branch of mechanics, supported by some of its most relevant results. We have attempted to attract a wide readership: we hope the book will be consulted not only by applied mathematicians, but also by mechanical, oil and chemical engineers, as well as those interested in flow through porous media or in the thermodynamic theory of mixtures.

In Chapter 1 the fundamental equations are derived by the homogenization process, the most important types of boundary condition are discussed. The use of dimensionless numbers to find the magnitude of various terms is demonstrated.

Some qualitative results on the problem of natural convection in bounded domains are proved in Chapter 2. The steady and non-steady cases, the structure of the solutions and the analyticity with respect to the Rayleigh number are studied in detail. In connection with the general results, the problem of convection in the domain consisting of the space between two concentric spheres is presented.

Chapter 3 is devoted to some important examples of convective motion in unbounded regions. The main types of boundary condition at infinity are presented; together with, a boundary-layer approximation for use at high Rayleigh numbers is presented. The many aspects not covered in this chapter are referenced in an extensive bibliography.

In Chapter 4 the governing equations of multiphase flow are deduced for the case of an arbitrary number of components in each phase, taking into account influences such as diffusion and chemical reactions. The complicated form of these equations may appear somewhat daunting; however, for single-phase flow they reduce to a system identical to that obtained in Chapter 1.

We discuss the equations which should be used for describing underground combustion in Chapter 5. The simplifying assumptions which it is possible to adopt cannot be treated exhaustively because a complete solution should appeal to the laws of physics and chemistry specific to each technique of combustion. Nevertheless, three examples which seem to us important are included. The reader interested in more details is invited to consult the works cited in the bibliography.

Finally, in two Appendices are presented the balance equations of continuous media, and notions and propositions from the theory of distribution and Sobolev spaces, which are necessary for an understanding of the mathematical analyses.

The notation and terminology used throughout are in essential agreement with those to be found in contemporary textbooks. Equations, statements and figures are numbered according to the section in which they appear; in the text the chapter number is included only if the equation, statement or figure being discussed is from a different chapter.

All the material in this book was originally presented at the seminar on 'Flow through Porous Media and Viscous Fluids', which is held at the University of Bucharest over the past twenty years; we should like to thank our seminar colleagues for the valuable discussions. Although some material has been published in research papers, most of it has never appeared in book form, while some has not been published in any form.

No systematic effort has been made to attribute results or to assign historical priorities.

Last but not least, we are grateful to our families for their constant support over many years.

*Bucharest, November 1985*

HORIA I. ENE
DAN POLIŠEVSKI

PART I

# Single-Phase Flow

CHAPTER 1

# The Homogenization Method for the Study of Fluid Flow in Porous Media

## 1. Homogenization of Second-Order Equation

The homogenization method is a mathematical method used for the study of the initial-value and boundary-value problems in domains with a periodic structure. If the dimensions of a period are small compared to the size of the whole domain, it is natural to define a small parameter $\varepsilon$ as the ratio between those dimensions. The study of the initial-value and boundary-value problems is associated with this small parameter. The asymptotic process $\varepsilon \to 0$ represents the transition from microscopic to macroscopic phenomena.

From the mathematical point of view the above problems consist in the study of the convergence of the solution of a family of partial differential operators, depending on the small parameter $\varepsilon$. Generally, this solution converges to the solution of another partial differential operator, called the homogenized operator. This homogenized operator describes the macroscopic behaviour of our medium, or more precisely the solution of the homogenized operator approximates the exact solution defined at the microscopic level, which is very difficult to calculate.

In order to explain this method, we consider the simplest boundary-value problem:

$$-\frac{\partial}{\partial x_i}\left(a_{ij}^{\varepsilon}(x)\frac{\partial u^{\varepsilon}}{\partial x_j}\right) = f(x) \quad \text{in}\,\Omega, \tag{1.1}$$

$$u^{\varepsilon}\,|_{\partial\Omega} = 0, \tag{1.2}$$

where $f(x)$ is a given smooth function defined on $\Omega$ and $a_{ij}^{\varepsilon}(x) = a_{ij}(x/\varepsilon)$ ($\varepsilon$ being a real positive parameter) are smooth real functions, $Y$-periodic, such that there exists $\gamma > 0$ with $a_{ij}(y)\xi_i\xi_j \geqslant \gamma\xi_i\xi_i$, $\forall y \in Y$, $\xi \in \mathbb{R}^3$. Here $\Omega$ is a bounded domain of the space $\mathbb{R}^3$ of coordinates $x_i$, inhomogeneous in the sense that the coefficients $a_{ij}$ are functions of $x$; more precisely they are $Y$-periodic functions of the variable $y$. In fact we define $Y = \Pi_{j=0}^{3}\,]0, y_j^0[\, \subset \mathbb{R}^3$ and a function $f: \mathbb{R}^3 \to \mathbb{R}^3$ is said to be $Y$-periodic if it admits the period $y_j^0$ in the direction $y_j$ $(j = 1, 2, 3)$.

Note that for fixed $\varepsilon > 0$, $u^{\varepsilon}$ exists and is unique.

The first step is to search for an asymptotic expansion of $u^\varepsilon(x)$ as function of $\varepsilon$ for $\varepsilon \to 0$. We shall look for $u^\varepsilon(x)$ in the form of a two-scale expansion:

$$u^\varepsilon(x) = u^0(x,y) + \varepsilon u^1(x,y) + \varepsilon^2 u^2(x,y) + \cdots, \qquad y = \frac{x}{\varepsilon}, \tag{1.3}$$

where the functions $u^j(x,y)$ are $Y$-periodic in the variable $y$, $\forall x \in \Omega$.

The idea of this step is to insert (1.3) in Equation (1.1) and to identify powers of $\varepsilon$. In order to do this we consider first $x$ and $y$ as independent variables, and next we replace $y$ by $x/\varepsilon$. The derivations may be understood as

$$\frac{\mathrm{d}}{\mathrm{d}x_i} \to \frac{\partial}{\partial x_i} + \frac{1}{\varepsilon}\frac{\partial}{\partial y_i}. \tag{1.4}$$

The total dependence on $x$ of a function $u(x,y)$ is obtained directly and through the variable $y$.

Using (1.3) in (1.1), we have successively

$$-\frac{\partial}{\partial y_i}\left(a_{ij}(y)\frac{\partial u^0}{\partial y_j}\right) = 0, \tag{1.5}$$

$$-\frac{\partial}{\partial y_i}\left(a_{ij}(y)\frac{\partial u^1}{\partial y_j}\right) - \frac{\partial}{\partial y_i}\left(a_{ij}(y)\frac{\partial u^0}{\partial x_j}\right) - \frac{\partial}{\partial x_i}\left(a_{ij}(y)\frac{\partial u^0}{\partial y_j}\right) = 0, \tag{1.6}$$

$$-\frac{\partial}{\partial y_i}\left(a_{ij}(y)\frac{\partial u^2}{\partial y_j}\right) - \frac{\partial}{\partial y_i}\left(a_{ij}(y)\frac{\partial u^1}{\partial x_j}\right) -$$

$$-\frac{\partial}{\partial x_i}\left(a_{ij}(y)\frac{\partial u^1}{\partial y_j}\right) - \frac{\partial}{\partial x_i}\left(a_{ij}(y)\frac{\partial u^0}{\partial x_j}\right) = f\cdot \tag{1.7}$$

Consider the following lemma: (Bensoussan *et al.* (1978)). The equation

$$-\frac{\partial}{\partial y_i}\left(a_{ij}(y)\frac{\partial \varphi}{\partial y_j}\right) = F; \quad \varphi \text{ } Y\text{-periodic} \tag{1.8}$$

admits a unique solution (up to an additive constant), iff

$$\int_Y F(y)\,\mathrm{d}y = 0. \tag{1.9}$$

It is easy to observe that the only periodic solution of (1.5) is $u^0(x,y) = u^0(x)$ (note that $x$ is a parameter).

With this result Equation (1.6) takes the form

$$-\frac{\partial}{\partial y_i}\left(a_{ij}(y)\frac{\partial u^1}{\partial y_j}\right) = \frac{\partial u^0}{\partial x_j}\frac{\partial a_{ij}}{\partial y_i}(y). \tag{1.10}$$

We now consider the problem of finding a solution $u^1(y)$ of Equation (1.10) which is $Y$-periodic, $x$ being a parameter; the right-hand side of (1.10) is then known. It is necessary to note that the $Y$-periodicity of $u^1(y)$ plays the role of boundary conditions. By virtue of (1.8) and (1.9), Equation (1.10) has a unique solution because $\int_Y(\partial a_{ij}/\partial y_i)\,dy = 0$, as a consequence of the $Y$-periodicity of the coefficients $a_{ij}(y)$.

In order to obtain the solution of this local (or microscopic) problem it is necessary to introduce the space

$$H_y = \{u \in H^1(Y) \mid u \text{ is } Y\text{periodic}, m(u) = 0\}, \tag{1.11}$$

where $m(\cdot)$ is the mean operator defined by

$$m(\varphi) = \frac{1}{|Y|}\int_Y \varphi(y)\,dy,$$

$|Y|$ being the measure of $Y$. Endowed with scalar product $(u, v)_{H_y} = \int_Y \nabla u\, \nabla v\, dy$, $H_y$ is a Hilbert space.

Now, we denote by $\chi^k$ the unique solution of the problem: find $\chi^k \in H_y$, such that

$$\int_Y a_{ij}(y)\frac{\partial \chi^k}{\partial y_j}\frac{\partial v}{\partial y_i}\,dy = \int_Y \frac{\partial a_{ij}}{\partial y_i}\, v\,dy, \quad \forall\, v \in H_y, \tag{1.12}$$

then the weak solution of (1.10) is

$$u^1(x,y) = \frac{\partial u^0}{\partial x}\chi^k(y) + c(x), \tag{1.13}$$

the term $c(x)$ being the additive constant.

Using (1.13), we then have

$$a_{ij}(y)\left(\frac{\partial u^0}{\partial x_j} + \frac{\partial u^1}{\partial y_j}\right) = \left[a_{ik}(y) + a_{ij}(y)\frac{\partial \chi^k}{\partial y_j}\right]\frac{\partial u^0}{\partial x_k}, \tag{1.14}$$

and applying the mean operator to Equation (1.7) we obtain

$$-\frac{\partial}{\partial x_i}\left(a^h_{ik}\frac{\partial u^0}{\partial x_k}\right) = f, \tag{1.15}$$

where

$$a^h_{ik} = m\left[a_{ik}(y) + a_{ij}(y)\frac{\partial \chi^k}{\partial y_j}\right] = m(a_{ik}) + m\left[a_{ij}\frac{\partial \chi^k}{\partial y_j}\right]. \tag{1.16}$$

Equation (1.15) is the homogenized (or macroscopic) equation, and the coefficients $a_{ik}^h$ are the homogenized coefficients (they are well determined constants which depend on the local coefficients $a_{ij}(y)$).

It is interesting to observe that the homogenized coefficients defined by (1.16) are not simply the mean values of the microscopic coefficients. It is easy to prove (Bensoussan *et al.* (1978), Sanchez-Palencia (1980)) that they have the following properties:

$$a_{ik}^h = a_{ki}^h,$$

$$a_{ik}^h \xi_i \xi_k \geqslant |\gamma \xi_i \xi_i, \quad \gamma > 0, \quad \forall \xi \in \mathbb{R}^3.$$

Of course, the boundary condition for the problem (1.15) is

$$u^0 |_{\partial\Omega} = 0. \tag{1.17}$$

The first term of the expansion (1.3) is then determined as the solution of an elliptic equation (1.15), with the boundary condition (1.17) and the constant coefficients given by (1.16). The formal point in the first step is the form of the expansion (1.3).

Now, the second step is to prove that $u^\varepsilon$ converges to the function $u^0$.

We begin with a lemma:

LEMMA 1.1. *Let* $\varphi \in L^2(Y)$. *If we extend* $\varphi$ *by periodicity to* $\mathbb{R}^n$, *then* $\varphi(x/\varepsilon) \to m(\varphi)$ *in* $L^2(\Omega)$ *weakly.*

*Proof.* Let $\theta \in \mathscr{D}(\Omega)$. We denote by $\theta_\varepsilon^*$ the function which takes in each period $\varepsilon Y$ a constant value equal to the value of $\theta$ at the centre of the period. The functions $\theta$ and $\theta_\varepsilon^*$ are continued with zero value in the periods which are not contained in $\Omega$. We denote also by $\varepsilon Y$ the smaller reunion of $\varepsilon Y$ periods which contain $\Omega$.

First we observe that the function $\varphi(x/\varepsilon)$ is bounded in $L^2(\Omega)$:

$$\int_\Omega \varphi^2\left(\frac{x}{\varepsilon}\right) \mathrm{d}x \leqslant \int_{\varepsilon Y} \varphi^2\left(\frac{x}{\varepsilon}\right) \mathrm{d}x = \Sigma\, \varepsilon^n \int_Y \varphi^2(y) \mathrm{d}y = n(\varepsilon)\, \varepsilon^n\, M = \frac{|\varepsilon Y|}{|Y|} M,$$

where $n(\varepsilon)$ is the number of periods which contain $\Omega$.

Because $\mathscr{D}(\Omega)$ is dense in $L^2(\Omega)$, it suffices to prove that

$$\int_\Omega \varphi\left(\frac{x}{\varepsilon}\right) \theta(x) \mathrm{d}x \to m(\varphi) \int_\Omega \theta(x) \mathrm{d}x, \quad \forall \theta \in \mathscr{D}(\Omega). \tag{1.18}$$

To do this, we first observe that the diameter of an $\varepsilon Y$-period is of order $\varepsilon$, and that from the conditions imposed on $\theta$ we have

$$|\theta(x) - \theta_\varepsilon^*(x)| \leqslant K\varepsilon, \quad \forall x \in \varepsilon Y.$$

Because $\theta$ is smooth, then

$$\int_\Omega \varphi\left(\frac{x}{\varepsilon}\right)(\theta(x) - \theta_\varepsilon^*(x)) \mathrm{d}x \to 0, \quad \varepsilon \to 0. \tag{1.19}$$

We now obtain

$$\int_\Omega \varphi\left(\frac{x}{\varepsilon}\right)\theta_\varepsilon^*(x)\mathrm{d}x = \int_{\varepsilon Y}\varphi\left(\frac{x}{\varepsilon}\right)\theta_\varepsilon^*(x)\mathrm{d}x - \int_{\partial\varepsilon}\varphi\left(\frac{x}{\varepsilon}\right)\theta_\varepsilon^*(x)\mathrm{d}x =$$

$$= m(\varphi)\int_{\varepsilon Y}\theta_\varepsilon^*(x)\mathrm{d}x - \int_{\partial\varepsilon}\varphi\left(\frac{x}{\varepsilon}\right)\theta_\varepsilon^*(x)\mathrm{d}x \to m(\varphi)\int_\Omega \theta(x)\mathrm{d}x, \tag{1.20}$$

where $\partial\varepsilon$ is $\varepsilon Y\backslash\Omega$ and has a diameter smaller than $\varepsilon\sqrt{2}$.

From (1.19) and (1.20), we obtain (1.18).

THEOREM 1.1. *Under the preceding hypothesis, if $u^\varepsilon$ is the solution of* (1.1) *and* (1.2), *and $u^0$ the solution of* (1.15) *and* (1.17), *then*

$$u^\varepsilon \to u^0 \quad \text{in } H_0^1(\Omega) \text{ weakly.} \tag{1.21}$$

*Proof.* The variational formulation of (1.1) and (1.2) ((1.15) and (1.17)) is: find $u^\varepsilon \in H_0^1(\Omega)$ $(u^0 \in H_0^1(\Omega))$ such that

$$\int_\Omega a_{ij}^\varepsilon \frac{\partial u^\varepsilon}{\partial x_i}\frac{\partial v}{\partial x_j}\mathrm{d}x = \int_\Omega f v\,\mathrm{d}x, \quad \forall v \in H_0^1(\Omega), \tag{1.22}$$

$$\int_\Omega a_{ij}^h \frac{\partial u^0}{\partial x_i}\frac{\partial v}{\partial x_j}\mathrm{d}x = \int_\Omega f v\,\mathrm{d}x, \quad \forall v \in H_0^1(\Omega). \tag{1.23}$$

If we take $v = u^\varepsilon$ in (1.22) and use the ellipticity condition for $a_{ij}(y)$, we get

$$\gamma\int_\Omega \frac{\partial u^\varepsilon}{\partial x_i}\frac{\partial u^\varepsilon}{\partial x_i}\mathrm{d}x \leqslant \int_\Omega a_{ij}^\varepsilon\frac{\partial u^\varepsilon}{\partial x_i}\frac{\partial u^\varepsilon}{\partial x_j}\mathrm{d}x \leqslant C'\|u^\varepsilon\|_{L^2} \leqslant C''\|u^\varepsilon\|_{H_0^1}.$$

Now, it is known that the bilinear form

$$a(u, v) = \int_\Omega \frac{\partial u}{\partial x_i}\frac{\partial v}{\partial x_j}\mathrm{d}x$$

is coercive on $H_0^1$; moreover,

$$a(u,u) \geqslant \inf\left(\frac{1}{2}, \frac{\gamma}{2}\right)\|u\|_{H_0^1}^2.$$

Using this result, we have

$$\|u^\varepsilon\|_{H_0^1} \leqslant C,$$

where $C$ is a constant independent of $\varepsilon$. We can extract a sequence denoted also by $u^\varepsilon$, such that for $\varepsilon \to 0$

$$u^\varepsilon \to u^* \quad \text{in } H_0^1(\Omega) \text{ weakly.} \tag{1.24}$$

In order to finish the proof of the theorem it is necessary to see that, for any sequence as (1.24), $u^* = u^0$.

From (1.24) we see that the partial derivatives of $u^\varepsilon$ are bounded in $L^2(\Omega)$. Then we have

$$\sigma_i^\varepsilon(x) \equiv a_{ij}^\varepsilon(x)\frac{\partial u^\varepsilon}{\partial x_j}, \qquad \|\sigma_i^\varepsilon(x)\|_{L^2} \leqslant C. \tag{1.25}$$

Note that $a_{ij}(x/\varepsilon)$ are bounded smooth functions. We can again extract a sequence, denoted also by $\sigma_i^\varepsilon$, such that for $\varepsilon \to 0$

$$\sigma_i^\varepsilon(x) \to \sigma_i^*(x) \quad \text{in } L^2(\Omega) \text{ weakly.} \tag{1.26}$$

Passing at the limit in (1.22) we obtain, $\forall v \in H_0^1$ fixed,

$$\int_\Omega \sigma_i^* \frac{\partial v}{\partial x_i}\,\mathrm{d}x = \int_\Omega f v\,\mathrm{d}x. \tag{1.27}$$

If $\chi^k(y)$ is the solution of (1.12), we take

$$w_\varepsilon(x) \equiv x_k + \varepsilon\chi^k\left(\frac{x}{\varepsilon}\right) \tag{1.28}$$

and it is clear that

$$w_\varepsilon \to x_k \quad \text{in } L^2(\Omega) \text{ strongly,}$$

and that

$$\int_\Omega a_{ij}\left(\frac{x}{\varepsilon}\right)\frac{\partial w_\varepsilon}{\partial x_j}\frac{\partial v}{\partial x_i}\,\mathrm{d}x = 0, \quad \forall v \in H_0^1(\Omega). \tag{1.29}$$

In Equation (1.22) we take $v = \varphi w_\varepsilon$, $\varphi \in \mathscr{D}(\Omega)$, and obtain

$$\int_\Omega a_{ij}\left(\frac{x}{\varepsilon}\right)\frac{\partial u^\varepsilon}{\partial x_j}\left(\frac{\partial \varphi}{\partial x_i} w_\varepsilon + \varphi\frac{\partial w_\varepsilon}{\partial x_i}\right)\mathrm{d}x = \int_\Omega f\varphi w_\varepsilon\,\mathrm{d}x. \tag{1.30}$$

In (1.29) we take $v = \varphi u^\varepsilon$, and obtain

$$\int_\Omega a_{ij}\left(\frac{x}{\varepsilon}\right)\frac{\partial w_\varepsilon}{\partial x_j}\left(\frac{\partial \varphi}{\partial x_i} u^\varepsilon + \varphi\frac{\partial u^\varepsilon}{\partial x_i}\right)\mathrm{d}x = 0. \tag{1.31}$$

By subtracting (1.31) from (1.30), we have

$$\int_\Omega a_{ij}\left(\frac{x}{\varepsilon}\right)\left[\frac{\partial u^\varepsilon}{\partial x_j}\frac{\partial \varphi}{\partial x_i} w_\varepsilon - \frac{\partial w_\varepsilon}{\partial x_j}\frac{\partial \varphi}{\partial x_i} u^\varepsilon\right]\mathrm{d}x = \int_\Omega f\varphi w_\varepsilon\,\mathrm{d}x. \tag{1.32}$$

In Equation (1.32) $a_{ij}\left(\frac{x}{\varepsilon}\right)\frac{\partial u^\varepsilon}{\partial x_j}$ converge in $L^2(\Omega)$ weakly (see (1.26)), $\frac{\partial \varphi}{\partial x_i} w_\varepsilon$ converge to $\frac{\partial \varphi}{\partial x_i} x_k$ in $L^2(\Omega)$ strongly, $a_{ij}\left(\frac{x}{\varepsilon}\right)\frac{\partial w_\varepsilon}{\partial x_i}$ are $\varepsilon Y$-periodic and according to Lemma 1.1 converge in $L^2$ weakly to its mean value $a^h_{ik}$, and $\frac{\partial\varphi}{\partial x_j}u^\varepsilon$ converge to $\frac{\partial\varphi}{\partial x_j}\,u^*$ strongly (see (1.24)). Then for $\varepsilon \to 0$ we have

$$\int_\Omega (\sigma^*_j\, x_k - a^h_{jk}\, u^*)\,\frac{\partial\varphi}{\partial x_j}\,\mathrm{d}x = \int_\Omega f\varphi x_k \mathrm{d}x \tag{1.33}$$

From (1.22) with $v = \varphi x_k$, we have

$$\int_\Omega \sigma^*_i \left(\frac{\partial\varphi}{\partial x_i}\,x_k + \delta_{ik}\varphi\right)\mathrm{d}x = \int_\Omega f\varphi x_k \mathrm{d}x,$$

and (1.33) takes the form

$$-\frac{\partial}{\partial x_j}(\sigma^*_j x_k - a^h_{jk} u^*) = -\frac{\partial\sigma^*_j}{\partial x_j}\,x_k$$

or

$$\sigma^*_j = a^h_{jk}\frac{\partial u^*}{\partial x_j}\cdot \tag{1.34}$$

With (1.34), (1.27) shows that $u^* \in H^1_0(\Omega)$ satisfies the variational formulation (1.23) of the problem for $u^0$, and then, by the uniqueness of the solution, we have $u^* = u^0$.

This example is the simplest in the theory of homogenization. Of course, it is possible to study the evolution problems and variational inequalities using the same ideas.

We have showed this problem only to illustrate the principal ideas behind the homogenization method. More details concerning the mathematical problems are to be found in Bensoussan *et al.* (1978), while more physical applications as well as mathematical aspects are given by Sanchez-Palencia (1980).

## 2. Darcy's Law and Continuity Equation

The problem of flow through a porous medium may be treated by the homogenization method, but only for a periodic model of such a medium. Nevertheless, this method affords a rigorous mathematical proof of Darcy's law. For this reason we explain here in detail the use of the method. The homogenization method also permits us to obtain a detailed description of microscopic as well as macroscopic phenomena; for example, it is possible to prove the tensorial character of the permeability as well as its symmetry and positivity.

In this section and in Section 3 we present only the asymptotic expansions for the general case. The convergence of the homogenization process is examined in

Section 5, but only for incompressible flow. Note that compressible flow is an open problem.

We consider the flow of a Newtonian fluid in a fluid domain $\Omega_{\varepsilon f}$ formed by the cavities of a rigid porous solid defined in the following way. We consider the parallelepipedic period $Y$ of the space of variables $y_i$ consisting of a fluid part $Y_f$ and a solid part $Y_s$, separated by a smooth boundary. Moreover, the union of all the $Y_f$ parts and the union of all the $Y_s$ parts (denoted also by $Y_f$ and $Y_s$, respectively) are connected.

The simplest representation of such a medium is a porous material formed by tubes parallel to the three directions of the coordinate axis, with a non-empty intersection (Fig. 1).

The pores thus have a periodic geometric structure, the period of which (associated with the small parameter $\varepsilon \to 0$) is small with respect to the other geometric parameters. We are still within the general framework of the homogenization method, but it is necessary to note that the difference from the problem presented in Section 1 is that the periodicity occurs in the geometry of the material, and not necessarily in the coefficients of the equations. Periodic coefficients do occur in the temperature equation, which is considered in Section 3. For the equations of conservation of mass and momentum, only the geometry of the medium plays an important role.

The asymptotic process and the limit equations may be very different in structure if several small parameters are involved in the problem. For example, if the viscosity coefficient is small, the velocity vector may be large, and the non-linear terms of the Navier–Stokes equation may appear in the asymptotic equation leading to a non-linear form of Darcy's law. Consequently we may find a very different asymptotic process if the given constants or data of the problem are considered as small parameters. The physical complications of the different results obtained when using small parameters are examined in Section 7 by introducing dimensionless numbers.

Now, if $\Omega$ is the porous body in the space of the variables $x_i$, we introduce the

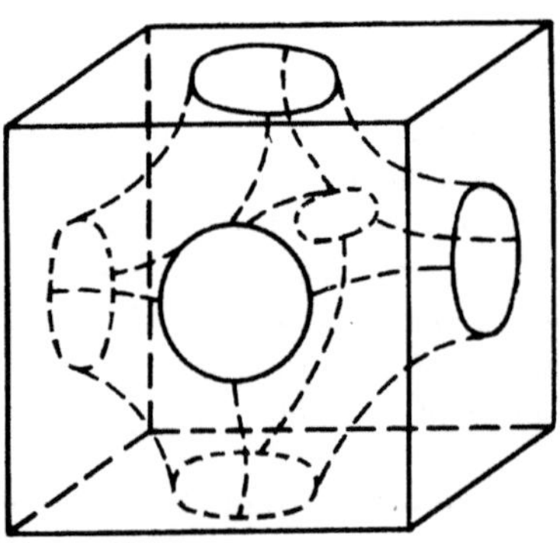

Fig. 1. Representation of a porous medium.

small positive parameter $\varepsilon$, and the fluid and solid domains $\Omega_{\varepsilon f}$ and $\Omega_{\varepsilon s}$:

$$\Omega_{\varepsilon f} = \{x \in \Omega;\, x \in \varepsilon Y_f\}, \qquad \Omega_{\varepsilon s} = \{x \in \Omega;\, x \in \varepsilon Y_s\}\,. \tag{2.1}$$

If $\rho^\varepsilon$, $p^\varepsilon$ and $\mathbf{v}^\varepsilon$ denote the density, pressure and velocity of the flow, they must satisfy the equations of conservation of mass and momentum:

$$\frac{\partial}{\partial x_i}\left(\rho^\varepsilon v_i^\varepsilon\right) = 0, \tag{2.2}$$

$$\rho^\varepsilon v_k^\varepsilon \frac{\partial v_i^\varepsilon}{\partial x_k} = -\frac{\partial p^\varepsilon}{\partial x_i} + \frac{\partial \tau_{ik}^\varepsilon}{\partial x_k} + \rho^\varepsilon f_i \tag{2.3}$$

in $\Omega_{\varepsilon f}$, where $f_i$ are the components of the exterior body force per unit mass, and $\tau_{ik}^\varepsilon$ are the components of the viscous stress tensor:

$$\tau_{ik}^\varepsilon = \lambda' \,\delta_{ik} \frac{\partial v_j^\varepsilon}{\partial x_j} + \mu' \left(\frac{\partial v_i^\varepsilon}{\partial x_k} + \frac{\partial v_k^\varepsilon}{\partial x_i}\right), \tag{2.4}$$

where $\lambda'$ and $\mu'$ are the two viscosity coefficients. We must add a state equation for the fluid. In Section 3 we derive a state equation in linear form in order to make explicit the influences of the temperature. This form is acceptable for not very large intervals of temperature and pressure, as is the case for flow in porous media (for a detailed discussion see Muskat (1946) or Bear (1972)).

The boundary condition on $\Gamma$, the fluid-solid interface, is

$$v_i^\varepsilon = 0\,. \tag{2.5}$$

Moreover, in order to study the asymptotic process we must consider expansions of the form (1.3) for the velocity and the pressure:

$$v_i^\varepsilon(x) = \varepsilon^n v_i^0(x,y) + \varepsilon^{n+1} v_i^1(x,y) + \cdots, \tag{2.6}$$

$$p^\varepsilon(x) = p^0(x,y) + \varepsilon p^1(x,y) + \cdots, \tag{2.7}$$

where $y = x/\varepsilon$ and all the functions are considered to be $Y$-periodic with respect to the variable $y$. For the density we consider also an expansion of the form

$$\rho^\varepsilon(x) = \rho^0(x,y) + \varepsilon\rho^1(x,y) + \cdots, \tag{2.8}$$

where $\rho^0$ is a linear function of the first term of the expansions for pressure and temperature; it is explained in Section 3.

Using (2.6) and (2.9) in (2.2), we have at order $\varepsilon^{n-1}$ and $\varepsilon^n$

$$\frac{\partial}{\partial y_i}(\rho^0 v_i^0) = 0, \tag{2.9}$$

$$\frac{\partial}{\partial x_i}(\rho^0 v_i^0) + \frac{\partial}{\partial y_i}(\rho^0 v_i^1 + \rho^1 v_i^0) = 0\,. \tag{2.10}$$

As is usual in homogenization problems, $\rho^0$ does not depend on $y$; it is a function of $x$ only, like the first term in the expansion of pressure or temperature. Equation (2.9) then takes the form

$$\operatorname{div}_y \mathbf{v}^0 = 0. \tag{2.11}$$

If we apply the mean-value operator (1.13) to Equation (2.10), we obtain the macroscopic continuity equation

$$\operatorname{div}_x (\rho^0 \mathbf{u}) = 0, \qquad m(\mathbf{v}^0) = \mathbf{u}. \tag{2.12}$$

Note that $\mathbf{v}^k$ as functions of $y$ are defined on $Y_f$, but it is natural to extend them to $Y$ with zero value on $Y_s$. In fact, from (2.5) they are zero on $\Gamma$. The mean value of the second term of Equation (2.10) is zero. To show that it suffices to take the mean value of a term of the form $\operatorname{div}_y \mathbf{v}^k$; consider the equations

$$m\left(\operatorname{div}_y \mathbf{v}^k\right) = \frac{1}{|Y|} \int_{Y_f} \frac{\partial v_i^k}{\partial y_i} \mathrm{d}y = \frac{1}{|Y|} \int_{\partial Y_f} n_i v_i^k \mathrm{d}s = 0.$$

The surface integral is zero because $n_i v_i^k$ is zero on $\Gamma$, and the integrals on the parts of $\partial Y_f$ lying on $\partial Y$ (Fig. 1) cancel by periodicity.

The only small parameter considered so far has been $\varepsilon$. In practice $\varepsilon$ is small, but it does not tend to zero when we deal with a porous medium. However, it is natural to assume that the flow closely corresponds to the asymptotic process $\varepsilon \to 0$. In practical problems other parameters may take small values. We show the effects of this by considering different values of the viscosity coefficients, for instance $\lambda' = \lambda \varepsilon^m$, and $\mu' = \mu \varepsilon^m$ ($\lambda$ and $\mu$ being constants).

Equation (2.3) together with Equations (2.4), (2.6), (2.7) and (2.8), then yields

$$\varepsilon^{2n-1} \rho^0 v_k^0 \frac{\partial v_i^0}{\partial y_k} + \cdots = -\varepsilon^{-1} \frac{\partial p^0}{\partial y_i} - \frac{\partial p^0}{\partial x_i} - \frac{\partial p^1}{\partial y_i} - \cdots$$
$$+ \varepsilon^{n+m-2} \left(\mu \Delta_y v_i^0 + (\lambda + \mu) \frac{\partial}{\partial y_i} (\operatorname{div}_y \mathbf{v}^0)\right) + \cdots + \rho^0 f_i + \cdots \tag{2.13}$$

or using (2.11) as well,

$$\varepsilon^{2n-1} \rho^0 v_k^0 \frac{\partial v_i^0}{\partial y_k} + \cdots = -\varepsilon^{-1} \frac{\partial p^0}{\partial y_i} - \frac{\partial p^0}{\partial x_i} - \frac{\partial p^1}{\partial y_i} - \cdots$$
$$+ \varepsilon^{n+m-2} \mu \Delta_y v_i^0 + \cdots + \rho^0 f_i + \cdots . \tag{2.14}$$

We now start the study of this problem by choosing $n+m-2=0$, since the usual values for liquid flow are $n=2$ and $m=0$. Of course there are other possibilities, and we discuss them below (Section 3). Consequently, from (2.14) at order $\varepsilon^{-1}$ and $\varepsilon^0$ we have

$$\frac{\partial p^0}{\partial y_i} = 0, \tag{2.15}$$

$$0 = -\frac{\partial p^1}{\partial y_i} + \mu \Delta_y v_i^0 + \left( f_i \rho^0 - \frac{\partial p^0}{\partial x_i} \right). \tag{2.16}$$

From (2.15) it is easy to see that $p^0 = p^0(x)$.

The local problem for the study of flow through porous media is then the solution of Equations (2.11) and (2.16). In order to study this problem, we define the space of $Y$-periodic functions

$$V_y = \{\mathbf{u}; \mathbf{u} \in H^1(Y_f); \mathbf{u}|_\Gamma = 0, \operatorname{div}_y \mathbf{u} = 0, \mathbf{u} \text{ is } Y\text{-periodic}\}, \tag{2.17}$$

which is a Hilbert space, with the scalar product

$$(\mathbf{u} \cdot \mathbf{w})_{V_y} = \int_{Y_f} \frac{\partial u_i}{\partial y_k} \frac{\partial w_i}{\partial y_k} \mathrm{d}y. \tag{2.18}$$

We take a test function $\mathbf{w} \in V_y$, and multiply (2.16) by $w_i$. By integration over $Y_f$, we have

$$-\int_{Y_f} \frac{\partial p^1}{\partial y_i} w_i \,\mathrm{d}y = -\int_{Y_f} \frac{\partial}{\partial y_i}(p^1 w_i) \mathrm{d}y = = \int_{\partial Y_f} p^1 w_i n_i \,\mathrm{d}s = 0, \tag{2.19}$$

$$\mu \int_{Y_f} \Delta_y v_i^0 w_i \,\mathrm{d}y = \mu \int_{Y_f} \left[ \frac{\partial}{\partial y_k} \left( \frac{\partial v_i^0}{\partial y_k} w_i \right) - \frac{\partial v_i^0}{\partial y_k} \frac{\partial w_i}{\partial y_k} \right] \mathrm{d}y =$$

$$= \mu \int_{\partial Y_f} \frac{\partial v_i^0}{\partial y_k} w_i n_k \,\mathrm{d}y - \mu \int_{Y_f} \frac{\partial v_i^0}{\partial y_k} \frac{\partial w_i}{\partial y_k} \mathrm{d}y = -\mu \int_{Y_f} \frac{\partial v_i^0}{\partial y_k} \frac{\partial w_i}{\partial y_k} \mathrm{d}y. \tag{2.20}$$

Note that the integrals over $\partial Y_f$ are zero because $\mathbf{w}|_\Gamma = 0$ and because of the periodicity condition.

By using (2.18) we find that

$$(\mathbf{v}^0, \mathbf{w})_{V_y} = \frac{1}{\mu} \left( \rho^0 f_i - \frac{\partial p^0}{\partial x_i} \right) \int_{Y_f} \mathbf{w}_i \,\mathrm{d}y \quad \forall \mathbf{w} \in V_y. \tag{2.21}$$

In order to complete the statement of this equivalence, if $\mathbf{v}^0$ is an element of $V_y$ and satisfies (2.21), by integration it satisfies

$$\int_{Y_f} \left[ \mu \Delta_y v_i^0 + \left( f_i \rho^0 - \frac{\partial p^0}{\partial x_i} \right) \right] w_i \,\mathrm{d}y = 0. \tag{2.22}$$

Then there exists $p^1$ of class $L^2$ satisfying (2.16). The only difficulty now is to prove that $p^1$ is a periodic function. Of course grad$p^1$ is periodic, by (2.16). Comparison with (2.22) and the results of the integration of (2.16), after multiplication by $w_i$, give us

$$\int_{\partial Y_f} p^1 w_i n_i \mathrm{d}s = 0\,,$$

and for arbitrary values of $w_i\, n_i$ we conclude that $p^1$ is a periodic function.

The variational formulation of the local problem given by (2.11) and (2.10) is then to find $\mathbf{v}^0 \epsilon\, V_y$ satisfying (2.21). The existence and uniqueness of the solution is an immediate consequence of the Lax – Milgran theorem.

By the linearity of this problem, if $\mathbf{v}^i \epsilon\, V_y$ satisfies

$$(\mathbf{v}^i, \mathbf{w})_{V_y} = \int_{Y_f} w_i\, \mathrm{d}y, \quad \forall\, \mathbf{w} \epsilon\, V_y, \tag{2.23}$$

then we have

$$\mathbf{v}^0 = \frac{1}{\mu}\left(f_i\, \rho^0 - \frac{\partial p^0}{\partial x_i}\right) \mathbf{v}^i. \tag{2.24}$$

Now, if we apply the mean operator (1.13) to (2.24), we obtain

$$u_j = \frac{K_{ij}}{\mu}\left(f_i\, \rho^0 - \frac{\partial p^0}{\partial x_i}\right), \quad K_{ij} = m\,(v_j^i)\,. \tag{2.25}$$

The relation (2.25) is *Darcy's law*. The constant tensor $K_{ij}$, defined as the mean value of the component $j$ of the vector $\mathbf{v}^i$, depends only on the geometry of the period $Y$, and is the usual permeability tensor. It is a symmetric, positive-definite matrix.

This is immediately apparent if in (2.23) we take $\mathbf{w} = \mathbf{v}^j$, and in the equation for $\mathbf{v}^j$ we take $\mathbf{w} = \mathbf{v}^i$. By the symmetry of the scalar product, $v_i^j = v_j^i$ and $K_{ij} = K_{ji}$. However, $K_{ij}\xi_i\xi_j = \xi_i\,\xi_j\, m\,(v_j^i) = (\xi^i \mathbf{v}^i, \xi_j \mathbf{v}^j)_{V_y} = \|\xi_i \mathbf{v}^i\|_{V_y} \geqslant 0$. Then, if we multiply (2.23) by $\xi_i$ and take a test function such that $m\,(w_i) = \xi_i$, we have $(\xi_i \mathbf{v}^i, \mathbf{w})_{V_y} = $
$= \xi_i \int_{Y_f} w_i\, \mathrm{d}y = \xi_i\, \xi_i$ , and $\xi_i\, \mathbf{v}^i = 0 \Rightarrow \xi_i = 0$.

For incompressible flow the continuity equation (2.12) takes the form

$$\mathrm{div}_x\, \mathbf{u} = 0\,. \tag{2.26}$$

In this case we have an elliptic equation for the unknown $p^0(x)$. If $p^0(x)$ is obtained, the velocity field $\mathbf{u}(x)$ is given by Darcy's law. In order to obtain $p^0(x)$ we need boundary conditions for the elliptic equation; for an impervious boundary the boundary condition is (see Section 5)

$$u_i n_i = 0,$$

which implies the Neumann boundary condition

$$\frac{K_{ij}}{\mu}\left(f_j\, \rho^0 - \frac{\partial p^0}{\partial x_j}\right) n_i = 0.$$

Darcy's law in the classical form (2.25) was obtained by setting $n=2$ and $m=0$, but the condition $n+m-2=0$ may be satisfied also, for example, if we choose $n=\frac{1}{2}$ and $m=\frac{3}{2}$, representing low viscosity. Consequently the velocity is large, and in the equation equivalent to (2.16) there may appear an additional non-linear term:

$$\rho^0 v_k^0 \frac{\partial v_i^0}{\partial y_k} = -\frac{\partial p^1}{\partial y_i} + \mu \Delta_y v_i^0 + \left( f_i \rho^0 - \frac{\partial p^0}{\partial x_i} \right).$$

However, this is a steady-state Navier – Stokes equation well-studied by classical methods (see Temam (1977) or Ladyzhenskaya (1969)). This problem give us a non-linear equation for $\mathbf{v}^0$, and consequently we have a non-linear filtration equation (which means that Darcy's law is not a deterministic law).

The relation $n+m-2=0$ for small viscosities (for example, it is also possible to take $m=2$ and $n=0$, or other values) usually gives rise to a non-linear form of Darcy's law which is difficult to explain. For gas flow, the situation may be the same.

In conclusion, it may be said that only for the flow of viscous, compressible liquids does the classical form of Darcy's law hold.

## 3. Thermal Equation for Flow in Porous Media

In order to complete the system of equations, it is necessary to obtain the macroscopic equation for the temperature.

If $T^\varepsilon$ denotes the temperature, then we have

$$\rho^\varepsilon c_f v_k^\varepsilon \frac{\partial T^\varepsilon}{\partial x_k} = \tau_{jk}^\varepsilon \frac{\partial v_j^\varepsilon}{\partial x_k} + \frac{\partial}{\partial x_k}\left( \chi_f' \frac{\partial T^\varepsilon}{\partial x_k} \right) \tag{3.1}$$

in $\Omega_{\varepsilon f}$, and

$$0 = \frac{\partial}{\partial x_k}\left( \chi_s' \frac{\partial T^\varepsilon}{\partial x_k} \right) \tag{3.2}$$

in $\Omega_{\varepsilon s}$. Here $\tau_{jk}^\varepsilon$ is given by (2.4).

The state equation for the fluid is

$$\rho^\varepsilon = \rho_0 (1 - \alpha T^\varepsilon + \beta p^\varepsilon). \tag{3.3}$$

The boundary conditions on $\Gamma$ are

$$T^\varepsilon |_f = T^\varepsilon |_s , \tag{3.4}$$

$$\chi_f' \frac{\partial T^\varepsilon}{\partial n}\bigg|_f = \chi_s' \frac{\partial T^\varepsilon}{\partial n}\bigg|_s . \tag{3.5}$$

We assume that the thermal expansion coefficient $\alpha$ and the compressibility coefficient $\beta$ are of the order of unity.

In considering the thermal conductivity coefficients $\chi'_{\mathrm{f}}$ and $\chi'_{\mathrm{s}}$, it is important to discuss different cases. It is well-known that the usual accepted hypothesis for flow through porous media is that there is an instantaneous thermal exchange between the fluid and the solid. This assumption makes it possible to use just one equation for the temperature field at the macroscopic level. From the mathematical point of view, this means that $\chi'_{\mathrm{f}}$ and $\chi'_{\mathrm{s}}$ are of the same order of magnitude. We consider then the case where $\chi' = \varepsilon^p \chi$, with constant $\chi$ in the two phases.

Using (2.6) and (2.7), we consider the temperature in the form of the expansion

$$T^{\varepsilon}(x) = T^0(x,y) + \varepsilon T^1(x,y) + \cdots . \tag{3.6}$$

Substituting (2.7) and (3.6) in (3.3), we obtain the density in the form of the expansion

$$\rho^{\varepsilon}(x) = \rho^0(x,y) + \varepsilon\rho^1(x,y) + \cdots, \qquad 3.7)$$

where

$$\rho^0(x,y) = \rho_0[1 - \alpha T^0(x,y) + \beta p^0(x)], \tag{3.8}$$

$$\rho^1(x,y) = \rho_0[ - \alpha T^1(x,y) + \beta p^1(x,y)] . \tag{3.9}$$

Note that (3.7) is now the explicit form of (2.8).

First, using (3.6) in (3.4) and (3.5), we establish the boundary conditions as

$$T^0\Big|_{\mathrm{f}} = T^0\Big|_{\mathrm{s}}, \tag{3.10}$$

$$T^1\Big|_{\mathrm{f}} = T^1\Big|_{\mathrm{s}}, \tag{3.11}$$

$$\chi_{\mathrm{f}} n_i \frac{\partial T^0}{\partial y_i}\bigg|_{\mathrm{f}} = \chi_{\mathrm{s}} n_i \frac{\partial T^0}{\partial y_i}\bigg|_{\mathrm{s}}, \tag{3.12}$$

$$\chi_{\mathrm{f}} n_i \left(\frac{\partial T^0}{\partial x_i} + \frac{\partial T^1}{\partial y_i}\right)\bigg|_{\mathrm{f}} = \chi_{\mathrm{s}} n_i \left(\frac{\partial T^0}{\partial x_i} + \frac{\partial T^1}{\partial y_i}\right)\bigg|_{\mathrm{s}}, \tag{3.13}$$

$$\chi_{\mathrm{f}} n_i \left(\frac{\partial T^1}{\partial x_i} + \frac{\partial T^2}{\partial y_i}\right)\bigg|_{\mathrm{f}} = \chi_{\mathrm{s}} n_i \left(\frac{\partial T^1}{\partial x_i} + \frac{\partial T^2}{\partial y_i}\right)\bigg|_{\mathrm{s}} . \tag{3.14}$$

Then from (3.1) and (3.2),

$$\varepsilon^{n-1}(\rho^0)^2 c_{\mathrm{f}} v_k^0 \frac{\partial T^0}{\partial y_k} + \varepsilon^n c_{\mathrm{f}} \left((\rho^0)^2 v_k^0 \frac{\partial T^0}{\partial x_k} + (\rho^0)^2 v_k^0 \frac{\partial T^1}{\partial y_k} + \right.$$

$$+ (\rho^0)^2 v_k^1 \frac{\partial T^0}{\partial y_k} + 2\,\rho^0 \rho^1 v_k^0 \frac{\partial T^0}{\partial y_k} + \cdots \Big) =$$

$$= \varepsilon^{m+2n-2} \rho^0 \tau^0_{jky} \frac{\partial v_j^0}{\partial y_k} + \cdots + \varepsilon^{p-2} \rho^0 \operatorname{div}_y (\chi_f \operatorname{grad}_y T^0) +$$

$$+ \varepsilon^{p-1} [\rho^0 \operatorname{div}_x (\chi_f \operatorname{grad}_y T^0) + \rho^0 \operatorname{div}_y (\chi_f \operatorname{grad}_x T^0) +$$

$$+ \rho^0 \operatorname{div}_y (\chi_f \operatorname{grad}_y T^1) + \rho^1 \operatorname{div}_y (\chi_f \operatorname{grad}_y T^0)] +$$

$$+ \varepsilon^p [\rho^0 \operatorname{div}_x (\chi_f \operatorname{grad}_x T^0) + \rho^0 \operatorname{div}_x (\chi_f \operatorname{grad}_y T^1) +$$

$$+ \rho^0 \operatorname{div}_y (\chi_f \operatorname{grad}_x T^1) + \rho^0 \operatorname{div}_y (\chi_f \operatorname{grad}_y T^2) +$$

$$+ \rho^1 \operatorname{div}_x (\chi_f \operatorname{grad}_y T^0) + \rho^1 \operatorname{div}_y (\chi_f \operatorname{grad}_x T^0) +$$

$$+ \rho^1 \operatorname{div}_y (\chi_f \operatorname{grad}_y T^1) + \rho^2 \operatorname{div}_y (\chi_f \operatorname{grad}_y T^0)] + \cdots, \tag{3.15}$$

$$0 = \varepsilon^{p-2} \operatorname{div}_y (\chi_s \operatorname{grad}_y T^0) + \varepsilon^{p-1} [\operatorname{div}_x (\chi_s \operatorname{grad}_y T^0) +$$

$$+ \operatorname{div}_y (\chi_s \operatorname{grad}_x T^0) + \operatorname{div}_y (\chi_s \operatorname{grad}_y T^1] +$$

$$+ \varepsilon^p [\operatorname{div}_x (\chi_s \operatorname{grad}_x T^0) + \operatorname{div}_x (\chi_s \operatorname{grad}_y T^1) +$$

$$+ \operatorname{div}_y (\chi_s \operatorname{grad}_x T^1) + \operatorname{div}_y (\chi_s \operatorname{grad}_y T^2)] + \cdots, \tag{3.16}$$

where

$$\tau^0_{jky} = \lambda\, \delta_{ik} \frac{\partial v_j^0}{\partial y_j} + \mu \left( \frac{\partial v_i^0}{\partial y_k} + \frac{\partial v_k^0}{\partial y_i} \right).$$

If we consider the range of validity of Darcy's law, $n + m - 2 = 0$, it is easy to see that when $p < n$, the terms representing viscous dissipation and convection are negligibly small (in fact they are of the order of $\varepsilon^n$). In this case the thermal equation for flow through porous media will be the conduction equation.

From the practical point of view it is interesting that if the order of magnitude of the thermal conductivity coefficient is smaller than that of the velocity, then the temperature equation is the same as it would be if the fluid were at rest.

This conclusion is also valid for low densities. To see that, it is sufficient to remark that the thermal conductivity of the gas is in general much smaller than the conductivity of the solid. The relation (3.5) may be approximated by $\chi_s \left.\frac{\partial T^\varepsilon}{\partial n}\right|_\Gamma = 0$; i.e., by a Neumann condition for the temperature in the solid part. This is sufficient for a study of the temperature variation in $\Omega_{\varepsilon s}$, which is easily obtained, as in Section 1, and gives us $T^0(x)$ and $T^1(x, y)$, which are well determined in $\Omega_{\varepsilon s}$. Moreover, from the boundary condition (3.10) we see that $T^0(x)$ is the same in the regions $\Omega_{\varepsilon s}$ and $\Omega_{\varepsilon f}$, so that $T^0$ may be considered as known in $\Omega_{\varepsilon f}$. Note that this case is not in the range of validity of Darcy's law (see the end of Section 1.2).

We now return to the case $p = n$. We shall see that $T^0$ does not depend on $y$. Equations (3.15) and (3.16) at order $\varepsilon^{n-2}$ give

$$\frac{\partial}{\partial y_i}\left(\chi_{ij}(y)\frac{\partial T^0}{\partial y_j}\right) = 0, \tag{3.17}$$

where $\chi$ take the values $\chi_s$ and $\chi_f$ in $Y_s$ and $Y_f$ respectively. Moreover, from (3.10) and (3.12) this equation holds over the whole $Y$ in the sense of distributions, and from the $Y$-periodicity we obtain $T^0 = T^0(x)$. Consequently, from (3.8) we have $\rho^0 = \rho^0(x)$.

In the same way, at order $\varepsilon^{n-1}$ we obtain

$$\frac{\partial}{\partial y_i}\left[\chi_{ij}(y)\left(\frac{\partial T^0}{\partial x_j} + \frac{\partial T^1}{\partial y_j}\right)\right] = 0 \tag{3.18}$$

or

$$-\frac{\partial}{\partial y_i}\left(\chi_{ij}(y)\frac{\partial T^1}{\partial y_j}\right) = \frac{\partial T^0}{\partial x_j}\frac{\partial \chi_{ij}}{\partial y_i}. \tag{3.19}$$

This is the classical equation (1.10), from which

$$m\left[\chi_{ij}(y)\left(\frac{\partial T^0}{\partial x_j} + \frac{\partial T^1}{\partial y_j}\right)\right] = \chi^h_{ij}\frac{\partial T^0}{\partial x_j}, \tag{3.20}$$

$$\chi^h_{ij} = m\left[\chi_{ij}(y) + \chi_{ik}(y)\frac{\partial \theta^j}{\partial y_k}\right] = m(\chi_{ij}) + m\left(\chi_{ik}\frac{\partial \theta^j}{\partial y_k}\right), \tag{3.21}$$

$$T^1(x,y) = \theta^j(y)\frac{\partial T^0}{\partial x_j} + c(x), \tag{3.22}$$

where $\theta^j$ is the solution of the problem (1.12).

At order $\varepsilon^n$, taking into account (3.18) and (3.20), together with the boundary conditions (3.13) and (3.14) and the $Y$-periodicity, Equations (3.15) and (3.16) give

$$\rho^0 c_f v^0_k\left(\frac{\partial T^0}{\partial x_k} + \frac{\partial T^1}{\partial y_k}\right) = \frac{\partial}{\partial y_j}\left[\chi_{jk}(y)\left(\frac{\partial T^1}{\partial x_k} + \frac{\partial T^2}{\partial y_k}\right)\right] + \tau^0_{jky}\frac{\partial v^0_j}{\partial y_k} +$$

$$+ \frac{\partial}{\partial x_j}\left[\chi_{jk}(y)\left(\frac{\partial T^0}{\partial x_k} + \frac{\partial T^1}{\partial y_k}\right)\right]. \tag{3.23}$$

If we take the mean value of Equation (3.23), we obtain succesively

$$\int_Y \frac{\partial}{\partial y_i}\left[\chi_{ij}(y)\left(\frac{\partial T^1}{\partial x_j} + \frac{\partial T^2}{\partial y_j}\right)\right] dy = \int_{\partial Y} n_i\left[\chi_{ij}(y)\left(\frac{\partial T^1}{\partial x_j} + \frac{\partial T^2}{\partial y_j}\right)\right] ds = 0,$$

$$m\left\{\frac{\partial}{\partial x_i}\left[\chi_{ij}(y)\left(\frac{\partial T^0}{\partial x_j}+\frac{\partial T^1}{\partial y_j}\right)\right]\right\} = \frac{\partial}{\partial x_i}\left(\chi_{ij}^h\frac{\partial T^0}{\partial x_j}\right),$$

$$m\left(\tau^0_{jky}\frac{\partial v_j^0}{\partial y_k}\right) = \frac{\mu}{|Y|}\left[\int_Y\frac{\partial v_j^0}{\partial y_k}\frac{\partial v_j^0}{\partial y_k}\mathrm{d}y + \int_Y\frac{\partial v_k^0}{\partial y_j}\frac{\partial v_j^0}{\partial y_j}\mathrm{d}y\right] =$$

$$= \frac{\mu}{|Y|}\left[\int_Y\frac{\partial v_j^0}{\partial y_k}\frac{\partial v_j^0}{\partial y_k}\mathrm{d}y + \int_Y\frac{\partial}{\partial y_j}\left(v_k^0\frac{\partial v_j^0}{\partial y_k}\right)\mathrm{d}y\right] =$$

$$= \frac{\mu}{|Y|}\left[\int_Y\frac{\partial v_j^0}{\partial y_k}\frac{\partial v_j^0}{\partial y_k}\mathrm{d}y + \int_{\partial Y} n_j\, v_k^0\frac{\partial v_j^0}{\partial y_k}\mathrm{d}s\right] =$$

$$= \frac{\mu}{|Y|}\int_Y\frac{\partial v_j^0}{\partial y_k}\frac{\partial v_j^0}{\partial y_k}\mathrm{d}y = \mu\,(K^{-1})_{ji}\,u_i u_j\,,$$

$$m\left(\rho^0 c_f v_k^0\left(\frac{\partial T^0}{\partial x_k}+\frac{\partial T^1}{\partial y_k}\right)\right) = \rho^0 c_f u_k\frac{\partial T^0}{\partial x_k}.$$

The macroscopic equation for the temperature is then

$$\rho^0 c_f u_k\frac{\partial T^0}{\partial x_k} = \mu\,(K^{-1})_{kj}\,u_j u_k + \frac{\partial}{\partial x_j}\left(\chi_{jk}^h\frac{\partial T^0}{\partial x_k}\right). \qquad (3.24)$$

Note that the macroscopic state equation is

$$\rho^0 = \rho_0\,(1-\alpha T^0+\beta p^0). \qquad (3.25)$$

The complete system of equations for the flow of a viscous compressible fluid through a porous medium is given by Equations (2.12), (2.25), (3.24) and (3.25). In fact these equations describe underground combustion.

An important remark on non-steady flow is that if we use a timescale contraction $\tau = \varepsilon^n t$, all considerations concerning Darcy's law hold; new terms appear in Equations (2.12) and (3.24).

Equation (2.10) contains a term of the form $\dfrac{\partial\rho^0}{\partial\tau}$, the mean value of which is

$$m\left(\frac{\partial\rho^0}{\partial\tau}\right) = \frac{\partial}{\partial\tau}\left(\frac{1}{|Y|}\int_{Y_f}\rho^0\,\mathrm{d}y\right) = \frac{\partial(n\rho^0)}{\partial\tau},$$

where $n$ is the porosity of the medium, defined by $n = \dfrac{|Y_f|}{|Y|}$.

In equation (3.24) appears a term of the form

$$m(\rho c)\frac{\partial T^0}{\partial\tau}$$

or

$$\left[n\rho_f c_f + (1-n)\,\rho_s c_s\right] \frac{\partial T^0}{\partial \tau}\,.$$

## 4. Natural Convection

Natural convection in porous media is somewhat different from the previous one. In fact it is necessary to take into consideration the motion of a slightly compressible viscous fluid. The density is then a function of temperature only, and the thermal expansion coefficient is small. Instead of the system of Equations (2.2), (2.3), (3.1) and (3.3), we consider the following:

$$\frac{\partial}{\partial x_i}\left(\rho^\varepsilon v_i^\varepsilon\right) = 0, \tag{4.1}$$

$$\rho^\varepsilon v_k^\varepsilon \frac{\partial v_i^\varepsilon}{\partial x_k} = -\frac{\partial P^\varepsilon}{\partial x_i} + \mu' \frac{\partial^2 v_i^\varepsilon}{\partial x_k \partial x_k} + (\rho^\varepsilon - \rho_0)\, g\delta_{i3}\,, \tag{4.2}$$

$$\rho^\varepsilon = \rho_0\,(1-\alpha' T^\varepsilon), \tag{4.3}$$

$$p^\varepsilon = P^\varepsilon + \rho_0 g x_3, \tag{4.4}$$

$$\rho^\varepsilon c_f v_k^\varepsilon \frac{\partial T^\varepsilon}{\partial x_k} = \tau_{jk}^\varepsilon \frac{\partial v_j^\varepsilon}{\partial x_k} + \frac{\partial}{\partial x_k}\left(\chi'_f \frac{\partial T^\varepsilon}{\partial x_k}\right), \tag{4.5}$$

where $p^\varepsilon$ is the pressure and $P^\varepsilon$ the difference between $p^\varepsilon$ and the Archimedes pressure for the reference temperature. The coefficients are $\mu' = \mu\varepsilon^m$, $\chi' = \chi\varepsilon^p$ and $\alpha' = \alpha\varepsilon^r$ $(0 < r < 1)$.

Equations (4.2) and (4.3), together with the condition $0 < r < 1$, constitute the Boussinesq approximation. According to this approximation, instead of (3.7) we consider the pressure in the form of the expansion

$$P^\varepsilon(x) = \varepsilon^r P^0(x,y) + \varepsilon^{r+1} P^1(x,y) + \cdots\,. \tag{4.6}$$

For the velocity and temperature the expansions are as before ((2.6) and (3.6)). From (4.3), for the density we have

$$\rho^\varepsilon(x) = \rho_0\,(1-\alpha\varepsilon^r T^0 - \alpha\varepsilon^{r+1} T^1 + \cdots)\,. \tag{4.7}$$

Using the same computation as in Section 2, it is easy to see that for $n+m-2 = r$, Darcy's law holds:

$$u_i = -\frac{K_{ij}}{\mu}\left(\frac{\partial P^0}{\partial x_j} + \alpha\,\rho_0\,T^0\,g\,\delta_{i3}\right), \tag{4.8}$$

and the continuity equation is of the same form as for incompressible fluids:

$$\operatorname{div}_x \mathbf{u} = 0\,. \tag{4.9}$$

In Equation (3.15) the terms for viscosity are of order $\varepsilon^{n+r}$, and are consequently negligibly small compared to the convective terms, which are of order $\varepsilon^n$. Consequently, instead of (3.24) we obtain

$$\rho_0 \, c_f \, u_k \frac{\partial T^0}{\partial x_k} = \frac{\partial}{\partial x_i}\left(\chi_{ij}^h \frac{\partial T^0}{\partial x_j}\right), \tag{4.10}$$

with $\chi_{ij}^h$ defined by (3.21).

The system of Equations (4.8)–(4.10) is the classical system of equations for natural convection in porous media.

## 5. Convergence of the Homogenization Process

The Boussinesq system, which in a certain approximation governs the thermal flow of viscous fluids (Landau and Lifchitz (1971), Chap. V) has been considered in Sections 2 and 4, where the macroscopic equations for this phenomenon have been obtained by applying the multiple-scale method of the homogenization theory to periodic models of porous media. We shall now prove the convergence of the homogenization process in the case of incompressible flow. A convergence theorem for such a homogenization process (the Stokes equations together with Darcy's law) can be found in Tartar (1980), and so far as we know it has not been extended yet. Here we treat a more realistic periodic model of porous media; a three-dimensional model with connected phases and a biphasic boundary. Moreover, we avoid any symmetry hypothesis for the tensor of thermal diffusion.

In Section 5.1 we discuss the derivation of the variational formulation of the natural convection problem. We then prove the existence of weak solutions, and give some estimates, including a maximum principle. Thus we can demonstrate the uniqueness result, on which is based the homogenization process. The convergence theorem is proved in Section 5.4 by using a special restriction operator constructed in Section 5.3. Finally, some properties of the homogenized coefficients are studied.

### 5.1. THE BOUSSINESQ SYSTEM

Without loss of generality, we can assume that $Y = [0, \delta[^3$ for some $\delta > 0$. Let $\Gamma$ be a surface of class $C^2$ included in $Y$, which crosses orthogonally the boundary following some regular curves which are reproduced identically on opposite faces; $\Gamma$ separates $Y$ into two domains, $Y_s$ (the solid part) and $Y_f$ (the fluid part), with the property that on repeating $Y$ by periodicity, the union of all the fluid parts is connected in $\mathbb{R}^3$ and is of class $C^2$; this applies to the solid parts too. The origin of the coordinate system is set in a fluid sphere; thus all the corners of $Y$ are surrounded by fluid neighbourhoods. We assume also that if $\Gamma$ attains an edge of $\bar{Y}$, then the normal of $\Gamma$ in that point is the edge itself.

Let $\Omega$ be an open, connected, bounded set in $\mathbb{R}^3$, locally located on one side of the boundary $\partial\Omega$, a manifold of class $C^2$, composed of a finite number of connected components, and let $\phi : \mathbb{R} \to [0, 1[$ be the function which associates with any real number its fractional part; defining $\varphi : \mathbb{R}^3 \to Y$ by

$$\varphi(x_1, x_2, x_3) = \delta(\phi(x_1/\delta), \phi(x_2/\delta), \phi(x_3/\delta)),$$

we say that a function $f : \mathbb{R}^3 \to \mathbb{R}$ is $Y$-periodic if $f = f \circ \varphi$.

Now, for any $\varepsilon \in (0, 1)$ we define

$$\varphi^{\varepsilon}(x) = \varphi(x/\varepsilon),$$

$$\Omega_{\varepsilon f} = \{x \in \Omega \,|\, \varphi^{\varepsilon}(x) \in Y_f\} := \text{the fluid part of } \Omega ,$$

$$\Omega_{\varepsilon s} = \{x \in \Omega \,|\, \varphi^{\varepsilon}(x) \in Y_s\} := \text{the solid part of } \Omega ,$$

$$\Gamma_{\varepsilon} = \bar{\Omega}_{\varepsilon f} \cap \Omega_{\varepsilon s} ,$$

$$(\partial\Omega)_{\varepsilon f} = \bar{\Omega}_{\varepsilon f} \cap \partial\Omega .$$

Considering thermal flow in $\Omega_{\varepsilon f}$ and pure conduction in $\Omega_{\varepsilon s}$, then the velocity $\mathbf{u}^{\varepsilon}$, the pressure $p^{\varepsilon}$ and the temperature $T^{\varepsilon}$ have to satisfy in some way the following system:

$$\operatorname{div} \mathbf{u}^{\varepsilon} = 0 \quad \text{in } \Omega_{\varepsilon f} , \tag{5.1.1}$$

$$-\nu\Delta\mathbf{u}^{\varepsilon} + (\mathbf{u}^{\varepsilon}\nabla)\mathbf{u}^{\varepsilon} + \nabla p^{\varepsilon} = [1 - \alpha(T - T_r)]\,\mathbf{g} \quad \text{in } \Omega_{\varepsilon f} , \tag{5.1.2}$$

$$\operatorname{div}(\mathbf{a}^{\varepsilon}\nabla T^{\varepsilon}) = \mathbf{u}^{\varepsilon}\nabla T^{\varepsilon} \quad \text{in } \Omega_{\varepsilon f} , \tag{5.1.3}$$

$$\operatorname{div}(\mathbf{a}^{\varepsilon}\nabla T^{\varepsilon}) = 0 \quad \text{in } \Omega_{\varepsilon s} , \tag{5.1.4}$$

$$\mathbf{a}^{\varepsilon}\nabla = \sum_{j=1}^{3} a_{ij}^{\varepsilon} \frac{\partial}{\partial x_j} ,$$

with the boundary conditions

$$\mathbf{u}^{\varepsilon} = \mathbf{0} \quad \text{on } \partial(\Omega_{\varepsilon f}) = \Gamma_{\varepsilon} \cup (\partial\Omega)_{\varepsilon f} , \tag{5.1.5}$$

$$T^{\varepsilon} = \tau \quad \text{on} \quad \partial\Omega , \tag{5.1.6}$$

and the transmission conditions

$$[T^{\varepsilon}]_{\varepsilon} = 0 , \tag{5.1.7}$$

where $[\,\cdot\,]_{\varepsilon}$ := the jump across $\Gamma_{\varepsilon}$, and

$$[\mathbf{a}^{\varepsilon}\nabla T^{\varepsilon}]_{\varepsilon}\, \mathbf{n}_{\varepsilon} = 0 , \tag{5.1.8}$$

where ($\mathbf{n}_{\varepsilon}$ := the unit normal on $\Gamma_{\varepsilon}$, exterior to $\Omega_{\varepsilon f}$. Here $\alpha > 0$ and $\nu > 0$ are respectively, the volume coefficient of thermal expansion and the kinematic viscosity of the fluid, $\mathbf{g} \in \mathbf{L}^3(\Omega)$ is the acceleration due to gravity, $\tau \in H^{3/2}(\partial\Omega)$ is the non-uniform temperature of the boundary (if $\tau$ is uniform the problem is trivial) and $T_r > 0$ is a uniform reference temperature defined as

$$T_r = (\sup_{x \in \partial\Omega} \tau + \inf_{x \in \partial\Omega} \tau)/2 \cdot$$

From the subsequent maximum principle (5.2.5) and a priori estimates (5.2.6)−(5.2.8), we shall learn that, independently of $\mathbf{a}^\varepsilon$, the velocity $\mathbf{u}^\varepsilon$ is of order $\varepsilon^{-2}$, while the temperature $T^\varepsilon$ is of order 1. As we look for a model of natural convection, we must bear in mind that both sides of (5.1.3) have to be of the same order; consequently the tensor of thermal diffusion $\mathbf{a}^\varepsilon$ has to be of order $\varepsilon^{-2}$. That is why we assume (the physical interpretation of this assumption is given in Section 6, by dimensional analysis) that

$$\mathbf{a}^\varepsilon(x) = \varepsilon^2 \mathbf{a}(x/\varepsilon),$$

where $\mathbf{a} = (a_{ij}) \in \mathbf{L}^\infty(\mathbb{R}^3)$ is $Y$-periodic and there exists $a > 0$ such that

$$a_{ij}\,\xi_i\,\xi_j \geqslant a\xi_i^2, \quad \forall \xi_i \in \mathbb{R},\ i = 1,2,3.$$

As usual, the scalar products and norms in $L^2(\Omega)$, $H^m(\Omega)$ and $H_0^1(\Omega)$ are denoted respectively by

$$(u, v) = \int_\Omega u \cdot v\,\mathrm{d}x, \qquad |u| = (u,u)^{1/2},$$

$$((u, v))_m = \sum_{|j| \leqslant m} \left(\frac{\partial u}{\partial x_j}, \frac{\partial v}{\partial x_j}\right), \qquad \|u\|_m = ((u,u))_m^{1/2},$$

$$((u, v)) = (\nabla u, \nabla v), \qquad \|u\| = ((u,u))^{1/2},$$

and the norm in $L^p(\Omega)$ $(p \neq 2)$ is denoted by $|\cdot|_p$.

We use the same notations for the scalar products and norms of $\mathbf{L}^2(\Omega) = [L^2(\Omega)]^3$, $\mathbf{H}^m(\Omega)$, $\mathbf{H}_0^1(\Omega)$ and $\mathbf{L}^p(\Omega)$. To the corresponding notations associated with $\Omega_{\varepsilon f}$ (instead of $\Omega$) we attach the index $\varepsilon$.

In order to get homogeneous boundary conditions, we need the following lemma:

LEMMA 5.1.1. *For any $h > 0$ there exists an element $\tau_h \in H^2(\Omega)$ with the properties*

$$\tau_h = \tau - T_\Gamma, \quad on\ \partial\Omega, \tag{5.1.9}$$

$$|S \nabla \tau_h| \leqslant h \|S\|, \quad \forall S \in H_0^1(\Omega). \tag{5.1.10}$$

*Proof.* Let $\rho(x) = d(x, \partial\Omega) :=$ the distance from $x$ to $\partial\Omega$. For any $\varepsilon > 0$, consider

$$\Omega_\varepsilon = \{x \in \Omega \mid \rho(x) < 2\exp(-1/\varepsilon)\}$$

and $f_\varepsilon \in C^2(\bar{\Omega})$, the Hopf function, which has the following properties:

(a) $f_\varepsilon = 1$ in some neighbourhood (which depends on $\varepsilon$) of $\partial\Omega$,
(b) $f_\varepsilon = 0$ in $\Omega \setminus \Omega_\varepsilon$,
(c) $\left|\dfrac{\partial f_\varepsilon}{\partial x_i}(x)\right| \leqslant \varepsilon/\rho(x)$ in $\bar{\Omega}_\varepsilon$, $\forall i = 1,2,3.$

(*Warning.* Whenever real numbers are concerned, as for instance in (c), the ordinary modulus must not be confused with the norm in $L^2(\Omega)$.)

Let $L(\tau)$ be the element of $H^2(\Omega)$ obtained by lifting the trace of $\tau \in H^{3/2}(\partial\Omega)$ (see Lions and Magenes (1968), Chap. I). Denoting by

$$\tau_\varepsilon = (L(\tau) - T_r)f_\varepsilon$$

for any $S \in H_0^1(\Omega)$, we have

$$\left|S\nabla\tau_\varepsilon\right| \leqslant \left[2\int_\Omega S^2(L(\tau)-T_r)^2(\nabla f_\varepsilon)^2\,dx + 2\int_\Omega S^2 f_\varepsilon^2(\nabla L(\tau))^2\,dx\right]^{1/2} \leqslant$$

$$\leqslant \varepsilon\sqrt{6}\,|(L(\tau)-T_r)S/\rho| + \sqrt{2}\,|S\nabla L(\tau)|_{L^2(\Omega_\varepsilon)} \leqslant$$

$$\leqslant \varepsilon\sqrt{6}\left|L(\tau) - T_r\right|_\infty \left|S/\rho\right| + \sqrt{2}\,|S|_6\,\left|\nabla L(\tau)\right|_6 \operatorname{meas}^{1/6}(\Omega_\varepsilon),$$

where meas $(\Omega_\varepsilon)$ is the measure of $\Omega_\varepsilon$.

With the Hardy inequality (see Lions and Magenes (1968), Chap. I, Théorème 11.3), i.e.

$$\left|\frac{S}{\rho}\right| \leqslant c(\Omega)\|S\|,$$

and the classical Sobolev embedding theorems, it follows that

$$|S\nabla\tau_\varepsilon| \leqslant c(\tau,\Omega)\max\{\varepsilon, \operatorname{meas}^{1/6}(\Omega)\}\,\|S\|.$$

Because for any $h > 0$ we can choose $\varepsilon_h > 0$ satisfying

$$c(\tau,\Omega)\max\{\varepsilon_h, \operatorname{meas}^{1/6}(\Omega_{\varepsilon_h})\} < h.$$

Then, setting $\tau_h := \tau_{\varepsilon_h}$, the proof is completed. □

Setting $S^\varepsilon = T^\varepsilon - (\tau_h + T_r)$ (later we shall choose a suitable value for the parameter $h > 0$), the system (5.1.1)–(5.1.8) becomes:

$$\operatorname{div}\mathbf{u}^\varepsilon = 0 \quad \text{in } \Omega_{\varepsilon f}, \tag{5.1.11}$$

$$-\nu\Delta\mathbf{u}^\varepsilon + (\mathbf{u}^\varepsilon\nabla)\mathbf{u}^\varepsilon + \nabla p^\varepsilon = [1-\alpha(S^\varepsilon + \tau_h)]\,\mathbf{g} \quad \text{in } \Omega_{\varepsilon f}, \tag{5.1.12}$$

$$\operatorname{div}(\mathbf{a}^\varepsilon\nabla(S^\varepsilon + \tau_h)) = \mathbf{u}^\varepsilon\nabla(S^\varepsilon + \tau_h) \quad \text{in } \Omega_{\varepsilon f}, \tag{5.1.13}$$

$$\operatorname{div}(\mathbf{a}^\varepsilon\nabla(S^\varepsilon + \tau_h)) = 0 \quad \text{in } \Omega_{\varepsilon s}, \tag{5.1.14}$$

$$\mathbf{u}^\varepsilon = \mathbf{0} \quad \text{on } \partial(\Omega_{\varepsilon f}), \tag{5.1.15}$$

$$S^\varepsilon = 0 \quad \text{on } \partial\Omega, \tag{5.1.16}$$

$$\left[S^\varepsilon\right]_\varepsilon = 0 \quad \text{on } \Gamma_\varepsilon, \tag{5.1.17}$$

$$\left[\mathbf{a}^\varepsilon\nabla(S^\varepsilon + \tau_h)\right]_\varepsilon\cdot\mathbf{n}_\varepsilon = 0 \quad \text{on } \Gamma_\varepsilon. \tag{5.1.18}$$

Let $\mathbf{V}_f^\varepsilon$ be the closure in $H_0^1(\Omega_{\varepsilon f})$ of

$$\mathscr{V}(\Omega_{\varepsilon f}) = \{\mathbf{v} \in \mathscr{D}(\Omega_{\varepsilon f}) \mid \operatorname{div}\mathbf{v} = 0 \text{ in } \Omega_{\varepsilon f}\}\cdot$$

Thus $X_\varepsilon = \mathbf{V}_f \times H_0^1(\Omega)$ is a Hilbert space with the scalar product

$$((\mathbf{u},S),(\mathbf{v},T))_{X_\varepsilon} = ((\mathbf{u},\mathbf{v}))_\varepsilon + ((S,T)).$$

Defining the mapping $G_\varepsilon : X_\varepsilon \to X'_\varepsilon$ by

$$\langle G_\varepsilon(\mathbf{u},S),(\mathbf{v},T)\rangle = \nu((\mathbf{u},\mathbf{v}))_\varepsilon + ((\mathbf{u}\nabla)\mathbf{u} + \alpha S\mathbf{g},\mathbf{v})_\varepsilon +$$

$$+ \lambda(\mathbf{a}^\varepsilon\nabla S,\nabla T) + \lambda(\mathbf{u}, T\nabla(S+\tau_h))_\varepsilon, \quad \forall(\mathbf{v},T)\in X_\varepsilon, \tag{5.1.19}$$

where the coupling parameter $\lambda > 0$ is chosen for convenience, we arrive at the variational formulation of the problem (5.1.11)–(5.1.18): to find $(\mathbf{u}^\varepsilon, S^\varepsilon)\in X_\varepsilon$ such that

$$\langle G_\varepsilon(\mathbf{u}^\varepsilon, S^\varepsilon),(\mathbf{v},T)\rangle = (1-\alpha\tau_h,\mathbf{g}\cdot\mathbf{v}) - \lambda(\mathbf{a}^\varepsilon\nabla\tau_h,\nabla T), \quad \forall(\mathbf{v},T)\in X_\varepsilon. \tag{5.1.20}$$

It is clear that if $(\mathbf{u}^\varepsilon, S^\varepsilon, p^\varepsilon)$ is a smooth solution of (5.1.11)–(5.1.18), then $(\mathbf{u}^\varepsilon, S^\varepsilon)$ is a solution of (5.1.20).

Conversely, if $(\mathbf{u}^\varepsilon, S^\varepsilon)\in X_\varepsilon$ satisfies (5.1.20), then by choosing the test functions in a proper manner we get

$$\langle -\nu\Delta\mathbf{u}^\varepsilon + (\mathbf{u}^\varepsilon\nabla)\mathbf{u}^\varepsilon - [1-\alpha(S^\varepsilon+\tau_h)]\,\mathbf{g},\mathbf{v}\rangle = 0, \quad \forall\mathbf{v}\in\mathbf{V}^\varepsilon_{\mathrm{f}}, \tag{5.1.21}$$

$$(\mathbf{a}^\varepsilon\nabla(S+\tau_h),\nabla T) + (\mathbf{u}^\varepsilon, T\nabla(S^\varepsilon+\tau_h))_\varepsilon = 0, \quad \forall T\in H^1_0(\Omega). \tag{5.1.22}$$

Supposing, possibly for a sub-sequence of $\varepsilon\to 0$, that $\partial(\Omega_{\varepsilon\mathrm{f}})$ is Lipschitz, we can give the following characterizations of $\mathbf{V}^\varepsilon_{\mathrm{f}}$ (Temam (1977), Chap. I):

$$\mathbf{V}^\varepsilon_{\mathrm{f}} = \{\mathbf{v}\in\mathbf{H}^1_0(\Omega_{\varepsilon\mathrm{f}}) \mid \operatorname{div}\mathbf{v} = 0\}, \tag{5.1.23}$$

If $\mathbf{f}\in\mathbf{H}^{-1}(\Omega_{\varepsilon\mathrm{f}})$ satisfies $\langle\mathbf{f},\mathbf{v}\rangle = 0$ $\forall\mathbf{v}\in\mathbf{V}^\varepsilon_{\mathrm{f}}$ then $\exists p\in L^2(\Omega_{\varepsilon\mathrm{f}})$ such that $\mathbf{f} = \nabla p$, (5.1.24)

From (5.1.23) we get (5.1.15) in $H^{1/2}(\partial(\Omega_{\varepsilon\mathrm{f}}))$, and (5.1.11) in $L^2(\Omega_{\varepsilon\mathrm{f}})$. As $\nabla\mathbf{u}^\varepsilon\in\mathbf{H}^{-1}(\Omega_{\varepsilon\mathrm{f}})$, we have $(\mathbf{u}^\varepsilon\nabla)\mathbf{u}^\varepsilon\in\mathbf{L}^{3/2}(\Omega_{\varepsilon\mathrm{f}})$ and $[1-\alpha(S^\varepsilon+\tau_h)]\mathbf{g}\in\mathbf{L}^2(\Omega_{\varepsilon\mathrm{f}})$, according to (5.1.24) it follows that $\exists p^\varepsilon\in L^2(\Omega_{\varepsilon\mathrm{f}})$ (determined up to an additive constant) such that (5.1.12) is satisfied in $\mathbf{H}^{-1}(\Omega_{\varepsilon\mathrm{f}})$. Now, if we continue $\mathbf{u}^\varepsilon$ (initially defined on $\Omega_{\varepsilon\mathrm{f}}$) to $\Omega$ with zero out of $\Omega_{\varepsilon\mathrm{f}}$, then (5.1.22) becomes

$$\langle -\operatorname{div}[\mathbf{a}^\varepsilon\nabla(S^\varepsilon+\tau_h)] + \mathbf{u}^\varepsilon\nabla(S^\varepsilon+\tau_h), T\rangle = 0, \quad \forall T\in H^1_0(\Omega). \tag{5.1.25}$$

Because $\operatorname{div}[\mathbf{a}^\varepsilon\nabla(S^\varepsilon+\tau_h)]\in H^{-1}(\Omega)$ and $\mathbf{u}^\varepsilon\nabla(S^\varepsilon+\tau_h)\in L^{3/2}(\Omega)$, then (5.1.13) and (5.1.14) are satisfied in $H^{-1}(\Omega)$. Also, $S^\varepsilon\in H^1_0(\Omega)$ implies (5.1.16) and (5.1.17) in the usual trace senses. Unfortunately, in order to find the sense of (5.1.18) we have to make a digression, slight generalization of Theorem 1.2 in Temam (1977, Chap. I).

THEOREM 5.1.1. *Let $\Omega$ be an open bounded set of class $C^2$ in $\mathbb{R}^N$, and let $E(\Omega) = \{\mathbf{u}\in L^2(\Omega) \mid \operatorname{div}\mathbf{u}\in L^p(\Omega)\}$, where $p = 2N/(N+2)$ if $N\geqslant 3$ and $p>1$ if $N=2$. Then $E(\Omega)$ is a Banach space equipped with the norm $|\mathbf{u}|_E = |\mathbf{u}| + |\operatorname{div}\mathbf{u}|_p$. Also, there exists a linear continuous operator $\Gamma_\mathbf{n}\in\mathscr{L}(E(\Omega), H^{-1/2}(\Omega))$ such that*

$$\Gamma_\mathbf{n}\mathbf{u} = \text{the restriction of } \mathbf{u}\cdot\mathbf{n} \text{ to } \partial\Omega, \ \forall\mathbf{u}\in\mathscr{E}(\bar\Omega), \tag{5.1.26}$$

*where* $\mathbf{n}$ *is the unit outward normal on* $\partial\Omega$, *and* $\forall \mathbf{u} \in E(\Omega)$, $\forall S \in H^1(\Omega)$ *the generalized Stokes formula holds*:

$$\int_\Omega (\mathbf{u}\cdot\nabla S + S \operatorname{div} \mathbf{u})\, dx = \langle \Gamma_{\mathbf{n}}\mathbf{u}, \Gamma_0 S \rangle_{(H^{-1/2}, H^{1/2})(\partial\Omega)}, \tag{5.1.27}$$

*where* $\Gamma_0 \in \mathscr{L}(H^1(\Omega), H^{1/2}(\partial\Omega))$ *is the usual trace operator.*

*Proof.* Let $\{\mathbf{u}_n\}_n$ be a Cauchy sequence in $E(\Omega)$. Then, there exists $\mathbf{u} \in \mathbf{L}^2(\Omega)$ such that $\mathbf{u}_n \to \mathbf{u}$ strongly in $\mathbf{L}^2(\Omega)$, and there exists $v \in L^p(\Omega)$ such that $\operatorname{div} \mathbf{u}_n \to v$ strongly in $L^p(\Omega)$. Denoting by $\langle \cdot,\cdot \rangle$ the duality product of $H^{-1}(\Omega)$ with $H_0^1(\Omega)$, and identifying, as usual, $L^2(\Omega)$ with its dual, for any $S \in H_0^1(\Omega)$ we get

$$\langle \operatorname{div} \mathbf{u}_n, S\rangle = -(\mathbf{u}_n, \nabla S) \to (-\mathbf{u}, \nabla S) = \langle \operatorname{div} \mathbf{u}, S\rangle\,. \tag{5.1.28}$$

From (5.1.28) it follows also that

$$|\operatorname{div} \mathbf{u}_n|_{H^{-1}(\Omega)} \leqslant |\mathbf{u}_n| \leqslant \text{const.},$$

and as $H^{-1}(\Omega)$ is precompact for the weak topology ($H_0^1(\Omega)$ is separable!) we obtain the convergence

$$\langle \operatorname{div} \mathbf{u}_n, S\rangle \to \langle v, S\rangle\,, \quad \forall S \in H_0^1(\Omega)\,,$$

and hence $v = \operatorname{div} \mathbf{u}$. Thus we have proved that $E(\Omega)$ is complete.

Now, let $\tau \in H^{1/2}(\partial\Omega)$ and let $S \in H^1(\Omega)$ with $\Gamma_0 S = \tau$. For any $\mathbf{u} \in E(\Omega)$, following the Lions – Magenes method (Lions and Magenes (1968)), we define

$$F_{\mathbf{u}}(\tau) = \int_\Omega (\mathbf{u} \nabla S + S \operatorname{div} \mathbf{u})\, dx.$$

We check that $F_{\mathbf{u}}(\tau)$ is well defined: let $S^1, S^2 \in H^1(\Omega)$ with $\Gamma_0 S^1 = \Gamma_0 S^2 = \tau$. Setting $S = S^1 - S^2 \in H_0^1(\Omega)$, and taking into account that $\mathscr{D}(\Omega)$ is dense in $H_0^1(\Omega)$, let us consider a sequence $\{S_n\}_n \subseteq \mathscr{D}(\Omega)$ such that $S_n \to S$ strongly in $H_0^1(\Omega)$. As

$$\int_\Omega (\mathbf{u} \nabla S_n + S_n \operatorname{div} \mathbf{u})\, dx = 0,$$

then from the relation

$$\left| \int_\Omega (\mathbf{u} \nabla S + S \operatorname{div} \mathbf{u})\, dx - \int_\Omega (\mathbf{u} \nabla S_n + S_n \operatorname{div} \mathbf{u}) dx \right| \leqslant$$

$$\leqslant |S - S_n|_{p_1} |\operatorname{div} \mathbf{u}|_p + |\mathbf{u}|\, \|S - S_n\| \leqslant C_1 \|S - S_n\|\, |\mathbf{u}|_E \to 0\,,$$

where $p_1$ is given by $1/p + 1/p_1 = 1$, we obtain

$$\int_\Omega (\mathbf{u} \nabla S^1 + S^1 \operatorname{div} \mathbf{u}) dx = \int_\Omega (u \nabla S^2 + S^2 \operatorname{div} \mathbf{u}) dx.$$

Now, from the Sobolev inequalities, it follows that

$$|F_{\mathbf{u}}(\tau)| \leqslant |S|_{p_1} |\operatorname{div} \mathbf{u}|_p + |\mathbf{u}|\, |\nabla S| \leqslant$$

$$\leqslant C_2 \|S\|_1 |\mathbf{u}|_E \leqslant C_3 |\tau|_{H^{1/2}(\partial\Omega)} |\mathbf{u}|_E\,. \tag{5.1.29}$$

From (5.1.29), and from the linearity of the mapping,

$$\tau \to F_{\mathbf{u}}(\tau),$$

there exists $\Gamma_{\mathbf{n}}(\mathbf{u}) \in H^{-1/2}(\partial\Omega)$ such that

$$\langle \Gamma_{\mathbf{n}}(\mathbf{u}), \tau \rangle_{(H^{-1/2}, H^{1/2})(\partial\Omega)} = F_{\mathbf{u}}(\tau).$$

Returning to (5.1.29), we find that $\Gamma_{\mathbf{n}} \in \mathscr{L}(E(\Omega), H^{-1/2}(\partial\Omega))$. As $\Gamma_{\mathbf{n}}$ satisfies (5.1.27) by construction, it remains only to prove (5.1.26).

For $\mathbf{u} \in \mathscr{E}(\bar{\Omega})$ and $S \in \mathscr{E}(\Omega)$, we have

$$F_{\mathbf{u}}(\Gamma_0 S) = \int_{\Omega} \operatorname{div}(S\mathbf{u})dx = \int_{\partial\Omega} S(\mathbf{u}\cdot\mathbf{n})dx = <\mathbf{u}\cdot\mathbf{n}, \Gamma_0 S>_{(H^{-1/2}, H^{1/2})(\partial\Omega)} . \tag{5.1.30}$$

Since $\{\Gamma_0 S | S \in \mathscr{E}(\bar{\Omega})\}$ is a dense subset of $H^{1/2}(\partial\Omega)$, (5.1.30) yields, by continuity,

$$F_{\mathbf{u}}(\tau) = <\mathbf{u}\cdot\mathbf{n}, \tau>_{(H^{-1/2}, H^{1/2})(\partial\Omega)}, \quad \forall \tau \in H^{1/2}(\partial\Omega),$$

and hence $\Gamma_{\mathbf{n}}(\mathbf{u}) = \mathbf{u}\cdot\mathbf{n}$ when $\mathbf{u} \in \mathscr{E}(\bar{\Omega})$. □

Coming back to our problem, as $\mathbf{a}^{\varepsilon}\nabla(S^{\varepsilon} + \tau_h) \in \mathbf{L}^2(\Omega)$ and $\operatorname{div}[\mathbf{a}^{\varepsilon}\nabla(S^{\varepsilon} + \tau_h)] = \mathbf{u}^{\varepsilon}\nabla(S^{\varepsilon} + \tau_h) \in L^{3/2}(\Omega)$, it is clear from Theorem 5.1.1 that (5.1.18) is satisfied in the $E(\Omega)$ trace sense.

## 5.2. THE WEAK SOLUTIONS

In proving the existence theorem for the variational problem (5.1.20), we make use of the following result (Gossez (1966)):

THEOREM 5.2.1. *Let $X$ be a reflexive Banach space, and $G: X \to X'$ a continuous mapping between the corresponding weak topologies. If*

$$\lim_{|u|_X \to \infty} \frac{<Gu, u>}{|u|_X} = \infty,$$

*then $G$ is a surjection.*

Now we can prove

THEOREM 5.2.2. *The problem (5.1.20) has at least one solution.*

*Proof.* Taking into account that $(1 - \alpha\tau_h)\mathbf{g} \in L^2(\Omega_{\varepsilon f})$ and $\operatorname{div}(\mathbf{a}^{\varepsilon}\nabla\tau_h) \in H^{-1}(\Omega)$, we have only to prove that $G_{\varepsilon} : X_{\varepsilon} \to X'_{\varepsilon}$, defined by (5.1.19), satisfies the hypothesis of Theorem 5.2.1.

First, we check the weak continuity of $G_{\varepsilon}$. Let $(\mathbf{u}_k, S_k) \rightharpoonup (\mathbf{u}, S)$ weakly in $X_{\varepsilon}$. Then, for any $(\mathbf{v}, T) \in X_{\varepsilon}$, we have

$$| < G_{\varepsilon}(\mathbf{u}_k, S_k) - G_{\varepsilon}(\mathbf{u}, S), (\mathbf{v}, T) > | \leqslant \nu |(\mathbf{u}_k - \mathbf{u}, \mathbf{v})_{\varepsilon}| +$$

$$+ \alpha|\mathbf{g}| \cdot |S_k - S|_4 |\mathbf{v}|_{4,\varepsilon} + |\mathbf{u}_k - \mathbf{u}|_{4,\varepsilon} \|\mathbf{u}_k\|_{\varepsilon} |\mathbf{v}|_{4,\varepsilon} + |\mathbf{u}|_{4,\varepsilon} \|\mathbf{v}\|_{\varepsilon} |\mathbf{u}_k - \mathbf{u}|_{4,\varepsilon} +$$

$$+ \lambda |(\nabla(S_k - S), {}^t a^{\varepsilon} \nabla T)| + \lambda |\mathbf{u}_k|_{4,\varepsilon} \|T\| \cdot |S_k - S|_4 + \lambda |\mathbf{u}_k - \mathbf{u}|_{4,\varepsilon} \|T\| \cdot |S + \tau_h|_4 ,$$

and the property follows because $\{|\mathbf{u}_k|_{4,\varepsilon}\}_k$ and $\{\|\mathbf{u}_k\|_\varepsilon\}_k$ are bounded, and the embeddings of $\mathbf{L}^4(\Omega_{\varepsilon f})$ in $\mathbf{V}_f^\varepsilon$ and of $L^4(\Omega)$ in $H_0^1(\Omega)$ are compact.

Now we shall prove that $G_\varepsilon$ is coercive. This is the stage at which $\lambda$ and $h$ become better defined.

If we put $(\mathbf{v}, T) = (\mathbf{u}, S)$ in (5.1.19), then

$$< G_\varepsilon(\mathbf{u}, S), (\mathbf{u}, S)> \; = \nu\|\mathbf{u}\|_\varepsilon^2 + \alpha(S, \mathbf{gu})_\varepsilon + \lambda\varepsilon^2(\mathbf{a}\nabla S, \nabla S) +$$

$$+ \lambda(\mathbf{u}, S\nabla\tau_h)_\varepsilon \geqslant \nu\|\mathbf{u}\|_\varepsilon^2 + \lambda a\varepsilon^2\|S\|^2 - C_1(\alpha|g|_3 + \lambda h)|\mathbf{u}|\|S\|, \tag{5.2.1}$$

where we have used (5.1.10) and some Sobolev inequalities. It seems from (5.2.1) that $\lambda$ and $h$ will not be uniform with respect to $\varepsilon$ but Tartar (1980) has shown that

$$|\mathbf{u}|_\varepsilon \leqslant K_2\varepsilon\delta\|\mathbf{u}\|_\varepsilon, \quad \forall \mathbf{u}\in H_0^1(\Omega_{\varepsilon f}), \tag{5.2.2}$$

where $K_2$ is independent of $\varepsilon$, $\delta$ and $\mathbf{u}$; thus (5.2.1) becomes

$$< G_\varepsilon(\mathbf{u}, S), (\mathbf{u}, S)> \; \geqslant \nu\|\mathbf{u}\|_\varepsilon^2 + \lambda a\varepsilon^2\|S\|^2 - C_2(\alpha|\mathbf{g}|_3 + \lambda h)\varepsilon\|\mathbf{u}\|_\varepsilon\|S\|, \tag{5.2.3}$$

and in order to get the desired property of $G_\varepsilon$ we have to impose the condition

$$C_2^2(\alpha|\mathbf{g}|_3 + \lambda h)^2 < 4\nu a\lambda. \tag{5.2.4}$$

Obviously, (5.2.4) can be satisfied with a sufficiently small $h$ and suitable $\lambda$. □

The following weak maximum principle is formulated in terms of inequality in the sense of $H^1(\Omega)$. We start, then, by recalling this notion and some propositions, following Kinderlehrer and Stampacchia (1980, Chap. II).

Let $u\in H^1(\Omega)$ and $E\subseteq\bar{\Omega}$; we say that $u$ is non-negative on $E$ in the sense of $H^1(\Omega)$ or, briefly that, $u\geqslant 0$ on $E$ in $H^1(\Omega)$, if there exists a sequence $u_n \in W_\infty^{(1)}(\Omega)$ such that $u_n(x)\geqslant 0$ for $x\in E$ and $u_n \to u$ in $H^1(\Omega)$. Let $v\in H^1(\Omega)$; naturally, we say that $u \leqslant v$ on $E$ in $H^1(\Omega)$ if $v-u\geqslant 0$ on $E$ in $H^1(\Omega)$. As $v$ may be a constant, we define

$$\sup_E u = \inf\{m\in\mathbb{R} \mid u\leqslant m \text{ on } E \text{ in } H^1(\Omega)\}.$$

Also, for any $x\in\Omega$, we say that $u(x) > 0$ in (the sense of) $H^1(\Omega)$ if there exists $B(x)$, a neighbourhood of $x$, and $\varphi$, a function from $\mathring{W}_\infty^{(1)}(B(x))$ with $\varphi\geqslant 0$ and $\varphi(x) > 0$, such that $u\geqslant\varphi$ on $B(x)$ in $H^1(\Omega)$. Let us remark that the set

$$\{x\in\Omega \mid u(x) > 0 \text{ in } H^1(\Omega)\}$$

is open.

PROPOSITION 5.2.1. *If $u\geqslant 0$ on $E$ in $H^1(\Omega)$, then $u\geqslant 0$, a.e. on $E$.*

PROPOSITION 5.2.2. *If $\sup_{\partial\Omega} u < \infty$, then for any $M$, $M\geqslant\sup_{\partial\Omega} u$, we have*

$$\max\{u - M, 0\}\in H_0^1(\Omega),$$
$$\max\{u - M, 0\}\geqslant 0 \quad \text{on } \Omega \text{ in } H^1(\Omega)\cdot$$

PROPOSITION 5.2.3. *Let $u\in W_p^{(1)}(\Omega)$ $(p\geqslant 1)$; then $v = \max\{u, 0\}\in W_p^{(1)}(\Omega)$, and* *we have in the sense of distributions:*

$$\nabla v = \begin{cases} \nabla u & \text{in } \{x \in \Omega \mid u(x) > 0 \text{ in } H^1(\Omega)\} \\ 0 & \text{otherwise.} \end{cases}$$

We can pass now to our maximum principle:

THEOREM 5.2.3. *If* $(\mathbf{u}^\varepsilon, S^\varepsilon) \in X_\varepsilon$ *is a solution of the problem* (5.1.20), *then* $S^\varepsilon \in L^\infty(\Omega)$ *and*

$$|S^\varepsilon + \tau_h|_\infty \leqslant \beta = \frac{1}{2}(\sup_{\partial\Omega}\tau - \inf_{\partial\Omega}\tau)\cdot \tag{5.2.5}$$

*Proof.* As $-\beta \leqslant \tau_h(x) \leqslant \beta$ for a.a. $x \in \Omega$, from Propositions 5.2.2 and 5.2.3 it follows that

$$R := \max\{S^\varepsilon + \tau_h - \beta, 0\} \in H_0^1(\Omega),$$

$$\nabla R = \begin{cases} \nabla(S^\varepsilon + \tau_h) & R \neq 0, \\ 0 & R = 0\,. \end{cases}$$

Setting $T = R$ in (5.1.22) we obtain

$$a\varepsilon^2 \|R\|^2 \leqslant (\mathbf{a}^\varepsilon \nabla R, \nabla R) = (\mathbf{a}^\varepsilon \nabla (S^\varepsilon + \tau_h), \nabla R) =$$

$$= -(\mathbf{u}^\varepsilon, R\nabla(S^\varepsilon + \tau_h))_\varepsilon = -(\mathbf{u}^\varepsilon, R\nabla R)_\varepsilon = 0\,,$$

and hence $R = 0$ in $H_0^1(\Omega)$; that is, $S^\varepsilon + \tau_h \geqslant \beta$ on $\Omega$ in $H^1(\Omega)$. According to Proposition 5.2.1,

$$S^\varepsilon + \tau_h \leqslant \beta \quad \text{a.e. on } \Omega\,.$$

Analogously, with $R := \min\{S^\varepsilon + \tau_h + \beta, 0\}$ we get $S^\varepsilon + \tau_h \geqslant -\beta$ a.e. on $\Omega$. Thus (5.2.5) is proved, as is the whole theorem because $\tau_h \in H^2(\Omega) \subseteq L^\infty(\Omega)$. □

PROPOSITION 5.2.4. *If* $(\mathbf{u}^\varepsilon, S^\varepsilon) \in X_\varepsilon$ *is a solution of the problem* (5.1.20), then

$$\|u^\varepsilon\| \leqslant \varepsilon\delta C_0\,, \tag{5.2.6}$$

$$|u^\varepsilon| < \varepsilon^2\delta^2 C_0 K_2, \quad (K_2 \text{ from } (5.2.2)) \tag{5.2.7}$$

$$\|S^\varepsilon\| \leqslant K_0\delta^2 + K_1, \tag{5.2.8}$$

*where* $C_0$, $K_0$ *and* $K_1$ *are independent of* $\varepsilon$ *and* $\delta$.

*Proof.* Setting $\mathbf{v} = \mathbf{u}^\varepsilon$ and $T = 0$ in (5.1.20), we obtain

$$\nu \|\mathbf{u}^\varepsilon\|_\varepsilon^2 = (\mathbf{u}^\varepsilon \mathbf{g}, 1 - \alpha(S^\varepsilon + \tau_h))_\varepsilon \leqslant (1 + \alpha\beta)\,|\mathbf{g}|\,|\mathbf{u}^\varepsilon|\,,$$

and by using (5.2.2) successively we obtain (5.2.6) and (5.2.7).

Taking in (5.1.22) $T = S^\varepsilon$, we get

$$(\mathbf{a}^\varepsilon \nabla S^\varepsilon, \nabla S^\varepsilon) = (\mathbf{u}^\varepsilon, (S^\varepsilon + \tau_h)\nabla S^\varepsilon)_\varepsilon - (\mathbf{a}^\varepsilon \nabla\tau_h, \nabla S^\varepsilon)\,,$$

and therefore

$$\varepsilon^2 a\|S^\varepsilon\| \leqslant \beta|\mathbf{u}^\varepsilon|_\varepsilon + \varepsilon^2|\mathbf{a}|_\infty|\nabla\tau_h|\,.$$

Introducing of (5.2.7) here yields (5.2.8). □

We now need a slight generalization of (5.2.2):

LEMMA 5.2.1. *For any real r, $2 \leqslant r \leqslant 6$, we have*

$$|\mathbf{u}|_{r,\varepsilon} \leqslant (\varepsilon\delta)^{\frac{(6-r)}{2r}} \cdot K_r \|\mathbf{u}\|_\varepsilon, \quad \forall \mathbf{u} \in \mathbf{H}_0^1(\Omega_{\varepsilon f}), \tag{5.2.9}$$

where $K_r$ is independent of ε, δ and **u**.

*Proof.* If $\mathbf{v} \in \mathbf{H}^1(Y_f/\delta)$, with $\mathbf{v} = \mathbf{0}$ on $\Gamma/\delta$, then we have a Friedrichs inequality of the form

$$|\mathbf{v}|_{\mathbf{H}^1(Y_f/\delta)} \leqslant C_1 |\nabla \mathbf{v}|_{\mathbf{L}^2(Y_f/\delta)}.$$

Using also the Sobolev inequality

$$|\mathbf{v}|_{\mathbf{L}^r(Y_f/\delta)} \leqslant C_2 |\mathbf{v}|_{\mathbf{H}^1(Y_f/\delta)},$$

we can conclude that

$$|\mathbf{v}|_{\mathbf{L}^r(Y_f/\delta)} \leqslant C_3 |\nabla \mathbf{v}|_{\mathbf{L}^2(Y_f/\delta)}. \tag{5.2.10}$$

Noting that every ε-cube is of the form

$$\varepsilon Y^n = \prod_{i=1}^{3} [\varepsilon\delta n_i, \varepsilon\delta n_i + \varepsilon\delta[,$$

with $n = (n_1, n_2, n_3) \in \mathbb{Z}^3$, then the fluid parts of the ε-cubes which intersect Ω can be indexed according to

$$\mathbb{Z}_f^\varepsilon = \{n \in \mathbb{Z}^3 |\ \varepsilon Y_f^n \cap \Omega \neq \emptyset\}.$$

Now, for some $\mathbf{u} \in \mathbf{H}_0^1(\Omega_{\varepsilon f})$, let us evaluate the integral

$$I_\varepsilon = \int_{\Omega_{\varepsilon f}} |\mathbf{u}(x)|^r \mathrm{d}x = \sum_{n \in \mathbb{Z}_f^\varepsilon} \int_{\varepsilon Y_f^n} |\mathbf{u}(x)|^r \mathrm{d}x.$$

Making the changes of variables

$$x = \varepsilon\delta n + \varepsilon\delta y, \qquad y \in Y_f/\delta \tag{5.2.11}$$

and using (5.2.10) yields

$$I_\varepsilon = \sum_{n \in \mathbb{Z}_f^\varepsilon} (\varepsilon\delta)^3 \int_{Y_f/\delta} |\mathbf{u}(\varepsilon\delta n + \varepsilon\delta y)|^r \mathrm{d}y \leqslant$$

$$\leqslant C_3^r (\varepsilon\delta)^3 (\varepsilon\delta)^r \sum_{n \in \mathbb{Z}_f^\varepsilon} \left[ \int_{Y_f/\delta} |\nabla_x \mathbf{u}(\varepsilon\delta n + \varepsilon\delta y)|^2 \mathrm{d}y \right]^{r/2} =$$

$$= C_3^r (\varepsilon\delta)^{r+3} \sum_{n \in \mathbb{Z}_f^\varepsilon} \left[ \frac{1}{(\varepsilon\delta)^3} \int_{\varepsilon Y_f^n} |\nabla_x \mathbf{u}(x)|^2 \mathrm{d}x \right]^{r/2} \leqslant$$

$$\leqslant C_3^r (\varepsilon\delta)^{3 - r/2} \|\mathbf{u}\|_\varepsilon^r.$$

□

REMARK 5.2.1. If $(\mathbf{u}^\varepsilon, S^\varepsilon) \in X_\varepsilon$ is a solution of the problem (5.1.20), then

$$|\mathbf{u}^\varepsilon|_{r,\varepsilon} \leqslant (\varepsilon\delta)^{\frac{(6+r)}{2r}} C_0 K_r, \tag{5.2.12}$$

which is obviously obtained from (5.2.6) and (5.2.9).

THEOREM 5.2.4. *If* $\delta > 0$ *is sufficiently small, then the solution of the problem* (5.1.20) *is unique for any* $\varepsilon \in (0, 1)$.

*Proof.* Let $(\mathbf{u}_1, S_1)$ and $(\mathbf{u}_2, S_2)$ be solutions of (5.1.20). Introducing $\mathbf{u} = \mathbf{u}_1 - \mathbf{u}_2$ and $S = S_1 - S_2$, and subtracting the corresponding relations, we have

$$\nu\,((\mathbf{u}, \mathbf{v}))_\varepsilon + (\alpha S\mathbf{g} + (\mathbf{u}\nabla)\,\mathbf{u}_2, \mathbf{v}) + ((\mathbf{u}_1\nabla)\mathbf{u}, \mathbf{v}) = 0, \quad \forall \mathbf{v} \in \mathbf{V}_\mathrm{f}^\varepsilon, \tag{5.2.13}$$

$$(\mathbf{a}^\varepsilon \nabla S, \nabla T) + (\mathbf{u}, T\,\nabla(S_2 + \tau_h))_\varepsilon + (\mathbf{u}_1, T\nabla S) = 0, \quad \forall T \in H_0^1(\Omega)\,. \tag{5.2.14}$$

Taking $\mathbf{v} = \mathbf{u}$ in (5.2.13) and $T = S$ in (5.2.14) and with (5.2.6), (5.2.9) and (5.2.12), it is easy to obtain the estimations

$$\|\mathbf{u}\|_\varepsilon \leqslant \text{const.}\,(\varepsilon\delta\,\|S\| + \varepsilon^{3/2}\delta^{3/2}\,\|\mathbf{u}\|_\varepsilon), \tag{5.2.15}$$

$$a\varepsilon\,\|S\| \leqslant \beta\delta\,\|\mathbf{u}\|_\varepsilon K_2\,. \tag{5.2.16}$$

Introducing (5.2.16) in (5.2.15) and taking in account that $\varepsilon \in (0, 1)$, we have

$$\|\mathbf{u}\|_\varepsilon \leqslant \text{const.}\,(\delta^2 + \delta^{5/2})\,\|\mathbf{u}\|_\varepsilon\,. \tag{5.2.17}$$

Obviously, the theorem follows as soon as

$$\delta^2 + \delta^{5/2} < (\text{const.})^{-1}, \tag{5.2.18}$$

this relation being ensured by the hypothesis. □

As the parameter $\delta > 0$ (the initial size of the cube) has no significance for our main purpose, the homogenization process, we shall consider from now on that it was chosen a priori – satisfying (5.2.18). In this way we overcome the uniqueness problem of viscous thermal flow, which seemed at the outset to be the main difficulty.

### 5.3. A RESTRICTION OPERATOR

In studying the homogenization process of the problem (5.1.20), one is faced with the difficulty that the velocity $\mathbf{u}^\varepsilon$ and pressure $p^\varepsilon$ are defined only in $\Omega_{\varepsilon\mathrm{f}}$. While the velocity can be naturally continued with zero in $\Omega_{\varepsilon\mathrm{s}}$, the prolongation of the pressure to $\Omega$ is not so straightforward.

Such a prolongation has been constructed by Tartar (1980), in $L^2(\Omega)$, by transposing some special restriction operator from $\mathbf{H}_0^1(\Omega)$ to $\mathbf{H}_0^1(\Omega_{\varepsilon\mathrm{f}})$. Unfortunately, it holds only when $Y_\mathrm{s}$ is strictly contained in $Y$; that is, from the physical point of view, for two-dimensional flow. Also, Tartar defines $\Omega_{\varepsilon\mathrm{f}}$ as the domain obtained from $\Omega$ by selecting only those parts of $\varepsilon Y_\mathrm{s}$ which do not intersect $\partial\Omega$, and so the border is obviously single-phase.

Here we extend Tartar's construction to the geometry discussed at the beginning of this section, which is obviously a three-dimensional geometry with connected phases and a two-phase boundary. The consequence of making this extension is that the prolongation of the pressure is made in $L^{6/5}(\Omega)$.

Without loss of generality, we can assume that $Y = [0,1[^3$. Then letting $\Sigma^i$, $i \in \{1, 2, ..., 6\}$ be the side faces of $Y$, the main result of this section is obtained as follows:

THEOREM 5.3.1. *For any* $\varepsilon > 0$ *sufficiently small, there exists a restriction operator* $R_\varepsilon \in \mathscr{L}(\mathring{W}_6^{(1)}(\Omega), H_0^1(\Omega_{\varepsilon f}))$ *such that*

(a) *If* $\mathbf{u} \in \mathring{W}_6^{(1)}(\Omega_{\varepsilon f})$ *is continued with zero in* $\Omega \backslash \Omega_{\varepsilon f}$, *then* $R_\varepsilon \mathbf{u} = \mathbf{u}$.
(b) *If* $\mathbf{u} \in \mathring{W}_6^{(1)}(\Omega)$ *and* $\operatorname{div} \mathbf{u} = 0$, *then* $\operatorname{div}(\mathbf{R}_\varepsilon \mathbf{u}) = 0$.
(c) *For any* $\mathbf{u} \in \mathring{W}_6^{(1)}(\Omega)$ *the following estimations hold*:

$$|\mathbf{R}_\varepsilon \mathbf{u}|_{L^2(\Omega_{\varepsilon f})} \leqslant c(Y_f)(|\mathbf{u}|_{L^6(\Omega)} + \varepsilon |\nabla \mathbf{u}|_{L^6(\Omega)}), \tag{5.3.1}$$

$$|\nabla \mathbf{R}_\varepsilon \mathbf{u}|_{L^2(\Omega_{\varepsilon f})} \leqslant c(Y_f)(|\mathbf{u}|_{L^6(\Omega)}/\varepsilon + |\nabla \mathbf{u}|_{L^6(\Omega)}). \tag{5.3.2}$$

The proof is based on two lemmas.

LEMMA 5.3.1. *There exists* $f \in \mathscr{L}(W_6^{(1)}(Y), H^{1/2}(\partial Y_f))$ *such that*

(a) $f(\mathbf{u}) = \mathbf{0}$ *on* $\Gamma$,
(b) $f(\mathbf{u}) = \mathbf{u}$ *on* $\partial Y_f$ *if* $\mathbf{u} = \mathbf{0}$ *in* $Y_s$,
(c) $\displaystyle\int_{\Sigma_f^i} f(\mathbf{u}) \cdot \mathbf{v}\, d\sigma = \int_{\Sigma^i} \mathbf{u} \cdot \mathbf{v}\, d\sigma, \ \forall\ i \in \{1, 2, ..., 6\}$,

*where* $\Sigma_f^i = \Sigma^i \cap \bar{Y}_f$ and $\Sigma_s^i = \Sigma^i \cap \bar{Y}_s$, *and* $\mathbf{v}$ *denotes the unit outward normal on* $\Sigma^i$.

*Proof.* Let $\gamma = \Gamma \cap \Sigma^1$, where $\Sigma^1$ is for instance the face $x_1 = 0$, and let $\gamma_k$ be a connected component of $\gamma$. Let $\gamma_k'$ be the curve obtained by a uniform 'dilation' of $\gamma_k$ on the normal towards $\Sigma_f^1$, of thickness $\tau_k > 0$ sufficiently small, such that if we denote by $D_k$ the domain contained between $\gamma_k$ and $\gamma_k'$, then $D_k \subset \Sigma_f^1$ and $D_k \cap D_j = \emptyset$ if $k \neq j$ (see Fig. 2). Obviously, we take advantage of our restrictive assumption that if $\Gamma$ attains an edge of $Y$, then the normal of $\Gamma$ at that point is the edge itself.

The parametric equations of the regular curve $\gamma_k$ are

$$x_2 = g(s), \qquad x_3 = h(s),$$

where $g, h \in C^2$ and $(g')^2 + (h')^2 = 1$, $s$ being the arc length on $\gamma_k$. Let us denote by $\lambda_k$ the length of $\gamma_k$. We may now assume that $\tau_k$ is also so small that the transformation from $(x_2, x_3) \in D_k$ to $(s, n) \in I_k = (0, \lambda_k) \times (0, \tau_k)$, given by

$$x_2 = g(s) - nh'(s), \qquad x_3 = h(s) + ng'(s),$$

has a strictly positive Jacobian determinant;

$$\frac{\partial(x_2, x_3)}{\partial(s, n)} = 1 + n(g''(s)h'(s) - g'(s)h''(s)) > 0.$$

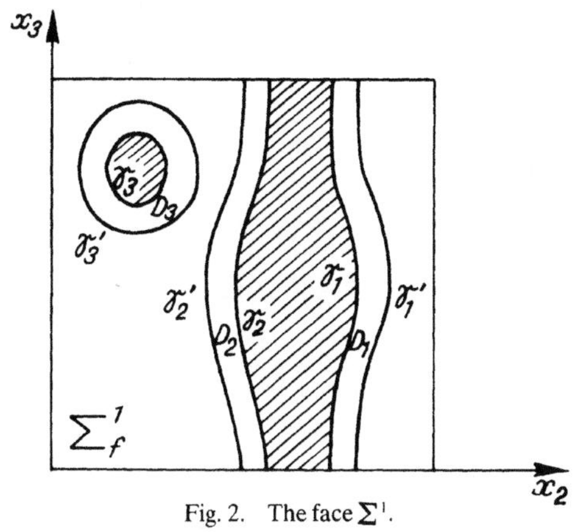


Fig. 2. The face $\Sigma^1$.

We then define in each $D_k$

$$f(\mathbf{u})(s,n) = \mathbf{u}(s,n) - P(n)\,\mathbf{u}(s,0), \quad s, n \in I_k\,,$$

where $P \in \mathbb{R}[X]$ satisfies the equations $P(\tau_k) = 0$, $P(0) = 1$ and

$$\int_0^{\tau_k} P(x)\,\mathrm{d}x = \int_0^{\tau_k} x P(x)\,\mathrm{d}x = 0\,.$$

In $\Sigma_f^1 \setminus (\bigcup_k D_k)$ we define

$$f(\mathbf{u}) = \mathbf{u} + \varphi \int_{\Sigma_f^1} \mathbf{u}\,\mathrm{d}x_2\mathrm{d}x_3\,,$$

where $\varphi \in \mathscr{D}(\Sigma_f^1 \setminus (\bigcup_k D_k))$ and $\int \varphi = 1$.

Taking into account that

$$\int_{D_k} f(\mathbf{u})\,\mathrm{d}x_2\mathrm{d}x_3 = \int_0^{\lambda_k}\int_0^{\tau_k} \mathbf{u}(s,n)\,\frac{\partial(x_2,x_3)}{\partial(s,n)}\,\mathrm{d}s\mathrm{d}n -$$
$$-\int_0^{\lambda_k}\int_0^{\tau_k} P(n)\mathbf{u}(s,0)\,\frac{\partial(x_2,x_3)}{\partial(s,n)}\mathrm{d}s\mathrm{d}n = \int_{D_k} \mathbf{u}\mathrm{d}x_2\mathrm{d}x_3\,,$$

the property (c) in Lemma 5.3.1 can easily be checked (for $i = 1$).

Using the same construction, we define $f(\mathbf{u})$ on the other side faces $\Sigma^i$. On $\Gamma$ we set $f(\mathbf{u}) = \mathbf{0}$, as property (a) requires.

We now have the entire definition of $f(\mathbf{u})$ on $\partial Y_f$. The fact that $\mathbf{u} = \mathbf{0}$ in $Y_s$ implies that $\mathbf{u}(s,0) = \mathbf{0}$, and also $f$ satisfies property (b). As $f$ is obviously linear, it remains to prove only the continuity. In fact we can go further than this and show that

$$|f(\mathbf{u})|_{W_6^{(1/2)}(\partial Y_f)} \leqslant c|\mathbf{u}|_{W_6^{(1)}(Y)}. \tag{5.3.3}$$

From the definition of $f$, we see that in order to prove (5.3.3) we have only to estimate the term $P(n)\mathbf{u}(s, 0)$ for $(s, n) \in I_k$:

$$|P(n)\mathbf{u}(s,0)|^6_{W_6^{(1/2)}(I_k)} = |P(n)\mathbf{u}(s,0)|^6_{L^6(I_k)} + \tag{5.3.4}$$

$$+\int_{I_k}\int_{I_k} \frac{|P(n_1)\mathbf{u}(s_1,0) - P(n_2)\mathbf{u}(s_2,0)|^6}{((s_1-s_2)^2 + (n_1-n_2)^2)^{5/2}}\, ds_1\, dn_1\, ds_2\, dn_2 \leqslant$$

$$\leqslant c_1|\mathbf{u}|^6_{L^6(\gamma_k)} + J_1 + J_2,$$

where

$$J_1 \leqslant \int_{I_k}\int_{I_k} \frac{|P(n_1) - P(n_2)|^6\, |\mathbf{u}(s_2,0)|^6}{|n_1-n_2|^5} \leqslant c_2\, |\mathbf{u}|^6_{L^\infty(\gamma_k)},$$

$$J_2 \leqslant \int_{I_k}\int_{I_k} \frac{|P(n_1)|^6|\mathbf{u}(s_1,0) - \mathbf{u}(s_2,0)|^6}{|s_1-s_2|^5} \leqslant c_3|\mathbf{u}|^6_{W_6^{(2/3)}(\gamma_k)}.$$

As $L^\infty(\gamma_k)$ is continuously embedded in $W_6^{(2/3)}(\gamma_k)$, from (5.3.4) we obtain

$$|P(n)\mathbf{u}(s,0)|_{W_6^{(1/2)}(I_k)} \leqslant C_4|\mathbf{u}|_{W_6^{(2/3)}(\gamma_k)}$$

and, using the continuity of the trace operator, the proof is completed. □

LEMMA 5.3.2. *If* $\mathbf{u} \in W_6^{(1)}(Y)$, *then there exist a unique* $\mathbf{v} \in H^1(Y_f)$ *and a unique* $\mathbf{q} \in L^2(Y_f)/R$, *solution of the problem*

$$-\Delta\mathbf{v} + \nabla q = -\Delta\mathbf{u} \quad in\ Y_f, \tag{5.3.5}$$

$$\operatorname{div}\mathbf{v} = \operatorname{div}\mathbf{u} + |Y_f|^{-1}\int_{Y_s} \operatorname{div}\mathbf{u}\, dy_s \quad in\ Y_f, \tag{5.3.6}$$

$$\mathbf{v} = f(\mathbf{u})\ on\ \partial Y_f \tag{5.3.7}$$

(*with f given by Lemma* 5.3.1). *Moreover, there exists a constant* $C(Y_f)$ *such that*

$$|\mathbf{v}|_{H^1(Y_f)} \leqslant C_1(Y_f)|\mathbf{u}|_{W_6^{(1)}(Y)}, \tag{5.3.8}$$

(*with* $|Y_f|$ *the measure of* $Y_f$).

*Proof.* Using the result obtained by Cattabriga (1961), we have only to check the compatibility condition

$$\int_{\partial Y_f} \mathbf{v}\cdot\nu\, d\sigma = \int_{Y_f} \operatorname{div}\mathbf{v}\, dy, \tag{5.3.9}$$

which follows from

$$\int_{\partial Y_f} f(\mathbf{u})\cdot\nu\, d\sigma = \sum_{i=1}^{6}\int_{\Sigma_f^i} f(\mathbf{u})\cdot\nu\, d\sigma = \sum_{i=1}^{6}\int_{\Sigma^i} \mathbf{u}\cdot\nu\, d\sigma =$$

$$= \int_{\partial Y} \mathbf{u}\cdot\mathbf{v}\, d\sigma = \int_Y (\operatorname{div}\mathbf{u} + |Y_f|^{-1}\int_{Y_s} \operatorname{div}\mathbf{u}\, dy_s)\, dy_f\,,$$

where we have used property (c) of Lemma 5.3.1 for $f$. □

*Proof of Theorem* 5.3.1. Noting that every $(\varepsilon Y)$ – cube is of the form

$$\varepsilon Y^n = \prod_{i=1}^{3} [\varepsilon n_i, \varepsilon n_i + \varepsilon[,$$

with $n = (n_1, n_2, n_3) \in \mathbb{Z}^3$, then the $(\varepsilon Y)$ – cube which intersect $\Omega$ can be indexed according to

$$\mathbb{Z}_\varepsilon = \{n \in \mathbb{Z}^3 \mid \varepsilon Y^n \cap \Omega \neq \emptyset\}.$$

For the $\varepsilon Y$-cubes which are cut by $\partial\Omega$, we shall denote by $(\tilde{\cdot})$ all the parts of them which are still contained in $\Omega$.

Recalling that $\partial\Omega$ is of class $C^2$, and that the corners of $Y$ are all surrounded by fluid, we take $\varepsilon > 0$ so small that if $\partial\Omega$ has an intersection with an $\varepsilon\tilde{\Sigma}^i$ side face, such that $|\varepsilon\tilde{\Sigma}^i_f| = 0$; then for the adjacent $\varepsilon Y$-cube outward to $\Omega$, $|\varepsilon\tilde{Y}_f| = 0$. This is always possible because of our assumption that the tangent plane to $\varepsilon\Gamma$ on $\varepsilon\Gamma \cap \varepsilon\Sigma^i$, if it exists, is orthogonal to $\varepsilon\Sigma^i$ (see Fig. 3).

Let $\mathbf{u} \in \mathring{W}_6^{(1)}(\Omega)$; in $\Omega_{\varepsilon s}$ we define

$$R_\varepsilon \mathbf{u} = 0. \tag{5.3.10}$$

In $\Omega_{\varepsilon f}$ we specify the construction of $R_\varepsilon$ in every $\varepsilon Y_f^n$, $n \in \mathbb{Z}_\varepsilon$. Therefore we have to consider two situations.

First, if $\varepsilon Y_n \subseteq \Omega$, then we define

$$R_\varepsilon \mathbf{u} = \mathbf{v} \circ \varphi^\varepsilon \quad \text{in } \varepsilon Y_f^n, \tag{5.3.11}$$

where $\mathbf{v}$ is given by Lemma 5.3.2 for $\mathbf{u}(\varepsilon n + \varepsilon(\cdot)) \in W_6^{(1)}(Y)$.

Second, if $\varepsilon Y^n$ is cut by $\partial\Omega$, then we can repeat the construction of $f$ on $\tilde{\Sigma}^i_f$ as in Lemma 5.3.1, but only if $|\tilde{\Sigma}^i_f| \neq 0$. Let $\tilde{\mathbf{u}} \in W_6^{(1)}(Y)$ be the continuation of $\mathbf{u}(\varepsilon n + \varepsilon(\cdot))$ with zero in $(Y \setminus \tilde{Y})$; we remark here that it makes sense since

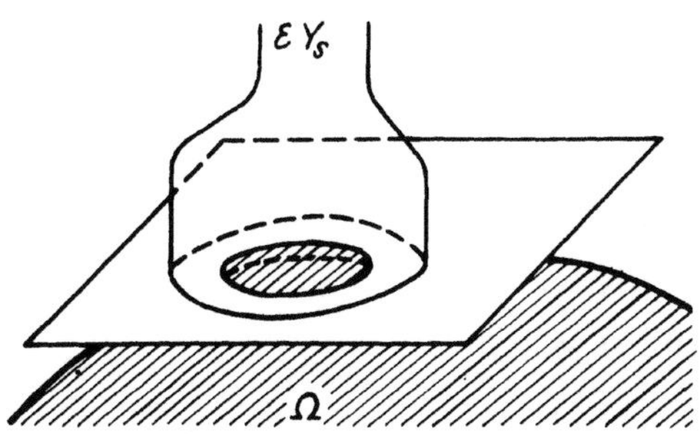


Fig. 3. The case $|\Sigma^i_f| = 0$.

$\mathbf{u} \in W_6^{(1)}(\Omega)$. Thus we can consider a system analogous to (5.3.5) – (5.3.7):

$$-\Delta \mathbf{v} + \nabla q = -\Delta \tilde{\mathbf{u}} \quad \text{in } Y_f, \tag{5.3.12}$$

$$\operatorname{div} \mathbf{v} = \operatorname{div} \tilde{\mathbf{u}} + |Y_f|^{-1} \int_{Y_s} \operatorname{div} \tilde{\mathbf{u}}\, dy_s - |Y_f|^{-1} \sum_i^{0} \int_{\tilde{\Sigma}_s^i} \tilde{\mathbf{u}} \cdot \mathbf{v}\, d\sigma \quad \text{in } Y_f, \tag{5.3.13}$$

$$\mathbf{v} = \mathrm{f}(\tilde{\mathbf{u}}) \quad \text{on } \tilde{\Sigma}_f, \qquad \mathbf{v} = 0 \quad \text{on } (\partial Y_f \setminus \tilde{\Sigma}_f'), \tag{5.3.14}$$

where $\Sigma_i^{o}$ denotes the sum extended over all those indices $i$ for which $|\tilde{\Sigma}_f^i| = 0$. Again, we have to check the compatibility relation (5.3.9):

$$\int_{Y_f} \operatorname{div} \mathbf{v} = \int_{\tilde{Y}} \operatorname{div} \tilde{\mathbf{u}} - \sum_i^{0} \int_{\tilde{\Sigma}_s^i} \tilde{\mathbf{u}} \cdot \mathbf{v}\, d\sigma = \sum_{i=1}^{6} \int_{\tilde{\Sigma}^i} \tilde{\mathbf{u}} \cdot \mathbf{v}\, d\sigma -$$

$$-\sum_i^{0} \int_{\tilde{\Sigma}_s^i} \tilde{\mathbf{u}} \cdot \mathbf{v}\, d\sigma = \sum_i^{1} \int_{\tilde{\Sigma}^i} \tilde{\mathbf{u}} \cdot \mathbf{v}\, d\sigma = \sum_i^{1} \int_{\tilde{\Sigma}_f^i} f(\tilde{\mathbf{u}}) \cdot \mathbf{v}\, d\sigma =$$

$$= \int_{\tilde{\Sigma}_f} f(\mathbf{u}) \cdot \mathbf{v}\, d\sigma,$$

where $\Sigma_i^1$ denotes the sum extended over all those indices $i$ for which $|\tilde{\Sigma}_f^i| \neq 0$. Hence, there exists a unique solution of (5.3.12) – (5.3.14), $\mathbf{v} \in H^1(Y_f)$ and $q \in L^2(Y_f)/\mathbb{R}$. If $\operatorname{div} \mathbf{u} = 0$ then, recalling the property which determined our choice of $\varepsilon$, in the outward adjacent $\varepsilon\tilde{Y}$-cube to a face with $|\varepsilon \tilde{\Sigma}_f^i| = 0$, we have

$$0 = \int_{\varepsilon \tilde{Y}} \operatorname{div} \mathbf{u} = \int_{\partial\Omega \cap \varepsilon \tilde{Y}} \mathbf{u} \cdot \mathbf{v}\, d\sigma + \int_{\varepsilon \tilde{\Sigma}_s^i} \mathbf{u} \cdot \mathbf{v}\, d\sigma$$

and, as $\mathbf{u} \in \mathring{W}_6^{(1)}(\Omega)$, it follows that

$$\int_{\tilde{\Sigma}_s^i} \tilde{\mathbf{u}} \cdot \mathbf{v}\, d\sigma = 0 .$$

Then, from (5.3.13), we get

$$\operatorname{div} \mathbf{v} = -|Y_f|^{-1} \sum_i^{0} \int_{\tilde{\Sigma}_s^i} \tilde{\mathbf{u}} \cdot \mathbf{v}\, d\sigma = 0 .$$

Moreover an estimation analogous to (5.3.8) holds:

$$|\mathbf{v}|_{H^1(Y_f)} \leqslant C_1(Y_f) |\tilde{\mathbf{u}}|_{W_6^{(1)}(Y)} . \tag{5.3.15}$$

But $\mathbf{v} = \mathbf{0}$ in $Y_f \setminus \tilde{Y}_f$, and thus (5.3.15) becomes

$$|\mathbf{v}|_{H^1(\tilde{Y}_f)} \leqslant C_1(Y_f) |\mathbf{u}(\varepsilon n + \varepsilon(\cdot))|_{W_6^{(1)}(\tilde{Y})} . \tag{5.3.16}$$

It is now obvious that in this final case we put

$$R_\varepsilon \mathbf{u} = \mathbf{v} \circ \varphi^{\varepsilon}, \tag{5.3.17}$$

with $\mathbf{v}$ given by (5.3.12)–(5.3.14). The properties (a) and (b) result straightforwardly from the definition of $R_\varepsilon$ in (5.3.10), (5.3.11) and (5.3.17). The property (c) can be proved by evaluating the integral

$$I_\varepsilon = \int_{\Omega_{\varepsilon f}} (|R_\varepsilon \mathbf{u}|^2 + \varepsilon^2 |\nabla R_\varepsilon \mathbf{u}|^2)\, dx .$$

We decompose $I_\varepsilon$ on every fluid part of the $\varepsilon Y^n$-cubes, $n \in \mathbb{Z}_\varepsilon$, and make the following change of variables:

$$x = \varepsilon n + \varepsilon y, \quad y \in Y. \tag{5.3.18}$$

Then, using for $\varepsilon Y^n$ either (5.3.8) or (5.3.16), and recomposing the right-hand side, via (5.3.18), we obtain

$$I_\varepsilon \leqslant C_1 \sum_{n \in \mathbb{Z}_\varepsilon} \varepsilon^2 \Big( \int_{\varepsilon Y^n} (|\mathbf{u}|^6 + \varepsilon^6 |\nabla \mathbf{u}|^6)\, dx \Big)^{1/3}. \tag{5.3.19}$$

With the classical Hölder inequality

$$\sum_{i=1}^{m} a_i^{1/3} \leqslant m^{2/3} \Big( \sum_{i=1}^{m} a_i \Big)^{1/3},$$

and taking in account that

$$\text{card}\, \mathbb{Z}_\varepsilon < 2\, |\Omega| / |\varepsilon Y| = C_2/\varepsilon^3 ,$$

from (5.3.19) we finally obtain

$$I_\varepsilon \leqslant C_1\, \varepsilon^2 (C_2\, \varepsilon^{-3})^{2/3} \Big( \sum_{n \in \mathbb{Z}_\varepsilon} \int_{\varepsilon Y^n} (|\mathbf{u}|^6 + \varepsilon^6 |\nabla \mathbf{u}|^6)\, dx \Big)^{1/3} \leqslant$$

$$\leqslant C_0 \big( |\mathbf{u}|^6_{L^6(\Omega)} + \varepsilon^6 |\nabla \mathbf{u}|^6_{L^6(\Omega)} \big)^{1/3},$$

which obviously completes the proof. □

## 5.4. THE CONVERGENCE THEOREM

For any $k \in \{1,2,3\}$, we consider two local-periodic problems.

PROBLEM 5.4.1. Find $\mathbf{u}^k \in V_Y$, given by (2.17), and $S^k \in H_Y$, given by (1.11), such that

$$\nu (\mathbf{u}^k, \mathbf{v})_V = \int_{Y_f} v_k\, dy, \quad \forall\, \mathbf{v} \in V_y , \tag{5.4.1}$$

$$\int_Y a_{ij} \frac{\partial S^k}{\partial y_i} \frac{\partial T}{\partial y_j}\, dy = - \int_Y a_{kj} \frac{\partial T}{\partial y_j}\, dy, \quad \forall T \in \tilde{H}^1_{\text{per}}(Y), \tag{5.4.2}$$

where $v_k$ is the $k$-component of $\mathbf{v}$.

Obviously, by the Lax-Milgram theorem there exists unique $\mathbf{u}^k \in V_Y$ and $S^k \in \tilde{H}^1_{\text{per}}(Y)$ satisfying (5.4.1) and (5.4.2) respectively. Also, from the regularity theorem of the Stokes problem (Temam (1977), Chap. I) we obtain $\mathbf{u}^k \in H^2(Y_f)$; moreover, using the flux property as used by Sanchez-Palencia (1980, Chap. 7), we can prove that there exists a unique $p^k \in H^1_{per}(Y_f)/\mathbb{R}$ such that

$$-\nu \Delta \mathbf{u}_k + \nabla p^k = \mathbf{e}^k \quad \text{in } Y_f, \tag{5.4.3}$$

$$\operatorname{div} \mathbf{u}^k = 0, \qquad \text{in } Y_f, \tag{5.4.4}$$

where $\mathbf{e}^k$ is the unit vector in the $k$-direction.

By methods similar to those used at the end of Section 5.1, we obtain

$$\operatorname{div}\left[{}^t\mathbf{a}\,\nabla(S^k + y_k)\right] = 0 \quad \text{in } H^{-1}(Y) \tag{5.4.5}$$

and, in the sense of $E(Y)$,

$$[{}^t\mathbf{a}\,\nabla(S^k + y_k)]_1 \cdot \mathbf{n} = 0 \quad \text{on } \Gamma, \tag{5.4.6}$$

where ${}^ta_{ij} = a_{ji}$, $[\cdot]_1$ is the jump across $\Gamma$, and $\mathbf{n}$ is the unit normal on $\Gamma$, exterior to $Y_f$.

If we introduce the subspace of $\mathbf{L}^2(\Omega)$,

$$\mathbf{H} = \{\mathbf{u} \in \mathbf{L}^2(\Omega) \mid \operatorname{div} \mathbf{u} = 0, \mathbf{u}\cdot\mathbf{v} = 0 \text{ on } \partial\Omega\},$$

we can now state our convergence theorem.

THEOREM 5.4.1. *If* $(\mathbf{u}^\varepsilon, S^\varepsilon, p^\varepsilon)$ *is the weak solution of the problem* (5.1.11)–(5.1.18), *and if we consider* $\mathbf{u}^\varepsilon$ *continued to* $\Omega$ *with value zero out of* $\Omega_{\varepsilon f}$, *then there exists a continuation of* $p^\varepsilon$ *to* $\Omega$ *(denoted by* $\tilde{p}^\varepsilon$*) such that*

$$\frac{1}{\varepsilon^2}\mathbf{u}^\varepsilon \rightharpoonup \mathbf{u} \quad \textit{weakly in } \mathbf{L}^2(\Omega), \tag{5.4.7}$$

$$S^\varepsilon \rightharpoonup S \qquad \textit{weakly in } H_0^1(\Omega), \tag{5.4.8}$$

$$\tilde{p}^\varepsilon \rightharpoonup p \qquad \textit{strongly in } L^{6/5}(\Omega), \tag{5.4.9}$$

*where* $(\mathbf{u}, S, p) \in \mathbf{H} \times H_0^1(\Omega) \times L^{6/5}(\Omega)/\mathbb{R}$ *satisfy 'weakly' in* $\Omega$ *the equations:*

$$u_i = K_{ij}\left\{-\frac{\partial p}{\partial x_j} + [1 - \alpha(S + \tau_h)]g_j\right\}, \tag{5.4.10}$$

$$-\operatorname{div}[\mathbf{A}\nabla(S + \tau_h)] + \mathbf{u}\nabla(S + \tau_h) = 0, \tag{5.4.11}$$

*the homogenized coefficient being defined by*

$$A_{ij} = m\left(a_{ij} + \frac{\partial S^i}{\partial Y_k} a_{kj}\right),$$

$$K_{ij} = \frac{1}{\mu(Y)} \int_{Y_f} u_j^i(y)\, dy,$$

$u_j^i$ *being the* $j$*-component of the vector* $\mathbf{u}^i$.

REMARK 5.4.1. The problem represented by (5.4.10) and (5.4.11) is studied in Chapter 2. Unless the corresponding dimensionless Rayleigh number is sufficiently small, the solution of this problem is probably not unique; that is why, for non-uniqueness, the convergences (5.4.7)−(5.4.9) hold only on some subsequences.

*Proof.* From (5.2.7) and (5.2.8) we can show that $\{\mathbf{u}^\varepsilon/\varepsilon^2\}_\varepsilon$ and $\{S^\varepsilon\}_\varepsilon$ are bounded in $\mathbf{L}^2(\Omega)$ and $H_0^1(\Omega)$ respectively; hence there exists $\mathbf{u} \in \mathbf{L}^2(\Omega)$ and $S \in H_0^1(\Omega)$ for which, passing, just in case, to a subsequence, the convergences (5.4.7) and (5.4.9) hold. Also, for any $q \in H^1(\Omega)$, with (5.1.27) we get

$$0 = (\mathbf{u}^\varepsilon, \nabla q) \to (\mathbf{u}, \nabla q)\,. \tag{5.4.12}$$

As the space $\mathbf{L}^2(\Omega)$ admits the orthogonal decomposition $\mathbf{L}^2(\Omega) = \mathbf{H} \oplus \mathbf{H}^\perp$ (Temam (1977), Chap. I), where

$$\mathbf{H}^\perp = \{\mathbf{u} \in \mathbf{L}^2(\Omega) \mid \exists q \in H^1(\Omega) \text{ such that } \mathbf{u} = \nabla q\},$$

from (5.4.14) it follows that $\mathbf{u} \in \mathbf{H}$.

We construct now the prolongation of $p^\varepsilon \in L^2(\Omega_{\varepsilon f})/\mathbb{R}$ defined by (5.1.24). As (5.1.12) is satisfied in $\mathbf{H}^{-1}(\Omega_{\varepsilon f})$, with the restriction operator of Theorem 5.3.1, for any $\mathbf{v} \in \mathring{W}_6^{(1)}(\Omega)$, we have

$$< \nabla p^\varepsilon, R_\varepsilon \mathbf{v} >_{(H^{-1}, H_0^1)(\Omega_{\varepsilon f})} = -\nu ((\mathbf{u}^\varepsilon, R_\varepsilon \mathbf{v}))_\varepsilon - ((\mathbf{u}^\varepsilon \nabla)\mathbf{u}^\varepsilon, R_\varepsilon \mathbf{v})_\varepsilon + \\ + (1 - \alpha(S^\varepsilon + \tau_h), \mathbf{g}\cdot R_\varepsilon \mathbf{v})\,. \tag{5.4.13}$$

Using (5.3.1) and (5.3.2), it follows that the functional

$$\mathbf{F}^\varepsilon(\cdot) = < \nabla p^\varepsilon, R_\varepsilon(\cdot) >$$

is bounded on $\mathring{W}_6^{(1)}(\Omega)$; that is, $\mathbf{F}^\varepsilon \in \mathbf{W}_{6/5}^{(-1)}(\Omega)$. If we continue $\mathbf{v} \in \mathring{\mathbf{W}}_6^{(1)}(\Omega_{\varepsilon f})$ with value zero in $\Omega_{\varepsilon s}$, from property (a) of Theorem 5.3.1 we have

$$\mathbf{F}^\varepsilon\big|_{\Omega_{\varepsilon f}} = \nabla p^\varepsilon. \tag{5.4.14}$$

Moreover, whenever div $\mathbf{v} = 0$, property (b) of Theorem 5.3.1 implies that

$$< \mathbf{F}^\varepsilon, \mathbf{u} >_{(\mathbf{W}_{6/5}^{(-1)}, \mathring{\mathbf{W}}_6^{(1)})(\Omega)} = 0\,.$$

Hence, $\exists \tilde{p}^\varepsilon \in \mathscr{D}'(\Omega)$ such that $\nabla \tilde{p}^\varepsilon = \mathbf{F}^\varepsilon \in \mathbf{W}_{6/5}^{(-1)}(\Omega)$. Using a result of Smith (1970) we get $\tilde{p}^\varepsilon \in L^{6/5}(\Omega)/\mathbb{R}$ and, comparing these results with (5.4.15), we see that $\tilde{p}^\varepsilon$ is the prolongation of $p^\varepsilon$. Furthermore, using (5.2.5) and some Hölder inequalities, we obtain from (5.4.13)

$$|< \nabla p^\varepsilon, \mathbf{v} >_{(\mathbf{W}_{6/5}^{(-1)}, \mathring{\mathbf{W}}_6^{(1)})(\Omega)}| \leqslant (\nu \|\mathbf{u}^\varepsilon\|_\varepsilon + |\mathbf{u}^\varepsilon|^2_{4,\varepsilon}) \|R_\varepsilon \mathbf{v}\| + \\ + (1 + \alpha\beta)|\mathbf{g}||R_\varepsilon \mathbf{v}|\,. \tag{5.4.15}$$

Recalling the estimates (5.2.6) and (5.2.12), and property (c) of Theorem 5.3.1 (that is, the inequalities (5.3.1) and (5.3.2)), the relation (5.4.15) yields

$$|< \nabla \tilde{p}, \mathbf{v} >_{(\mathbf{W}_{6/5}^{(-1)}, \mathring{\mathbf{W}}_6^{(1)})(\Omega)}| \leqslant \text{const.}\,(|\mathbf{v}|_6 + \varepsilon|\nabla \mathbf{v}|_6)\,.$$

Hence $|\nabla \tilde{p}^{\varepsilon}|_{\mathbf{W}_{6/5}^{(-1)}(\Omega)} \leqslant$ const., and consequently, with the inequality

$$|\tilde{p}^{\varepsilon}|_{L^{6/5}(\Omega)/R} \leqslant C(\Omega)|\nabla \tilde{p}^{\varepsilon}|_{\mathbf{W}_{6/5}^{(-1)}(\Omega)}, \tag{5.4.16}$$

which follows from Nečas (1966), we find that the sequence $\{\tilde{p}^{\varepsilon}\}_{\varepsilon}$ is bounded in $L^{6/5}(\Omega)/\mathbb{R}$, and therefore $\exists p \in L^{6/5}(\Omega)/\mathbb{R}$ such that on some sub-sequence

$$\tilde{p}^{\varepsilon} \rightharpoonup p \quad \text{weakly in } L^{6/5}(\Omega)/\mathbb{R},$$

$$\nabla \tilde{p}^{\varepsilon} \rightharpoonup \nabla p \quad \text{weakly in } \mathbf{W}_{6/5}^{(-1)}(\Omega).$$

Note that for any $\mathbf{w}^{\varepsilon} \rightharpoonup \mathbf{w}$ weakly in $\mathring{\mathbf{W}}_{6/5}^{(1)}(\Omega)$, we have

$$|<\nabla p^{\varepsilon}, \mathbf{w}^{\varepsilon}> - <\nabla p, \mathbf{w}>| \leqslant |<\nabla p^{\varepsilon}, \mathbf{w}^{\varepsilon} - \mathbf{w}>| + |<\nabla p^{\varepsilon} - \nabla p, \mathbf{w}>| \leqslant$$

$$\leqslant C(|\mathbf{w}^{\varepsilon} - \mathbf{w}|_6 + \varepsilon|\nabla(\mathbf{w}^{\varepsilon} - \mathbf{w})|_6) + (\text{terms which } \to 0).$$

As the embedding of $\mathring{\mathbf{W}}_6^{(1)}(\Omega)$ in $\mathbf{L}^6(\Omega)$ is compact, it follows that

$$<\nabla \tilde{p}^{\varepsilon}, \mathbf{w}^{\varepsilon}> \to <\nabla p, \mathbf{w}>,$$

that is, $\nabla \tilde{p}^{\varepsilon} \to \nabla p$ strongly in $\mathbf{W}_{6/5}^{(-1)}(\Omega)$. Finally, (5.4.9) can be obtained by recalling (5.4.16).

Now, it remains only to prove that $(\mathbf{u}, S, p)$ satisfy (5.4.10) and (5.4.11), to which end we apply a standard method.

We write the local equations (5.4.3)–(5.4.5) in terms of $x = \varepsilon y$, setting $\mathbf{v}^{\varepsilon}(x) = \mathbf{u}^k(x/\varepsilon)$, $T^{\varepsilon}(x) = x_k + \varepsilon S^k(x/\varepsilon)$ and $q_i^{\varepsilon}(x) = p^k(x/\varepsilon)$:

$$\operatorname{div} \mathbf{v}^{\varepsilon} = 0, \tag{5.4.17}$$

$$-\varepsilon^2 \nu \Delta \mathbf{v}^{\varepsilon} + \varepsilon \nabla q^{\varepsilon} = \mathbf{e}^k, \tag{5.4.18}$$

$$\operatorname{div}({}^t\mathbf{a} \nabla T^{\varepsilon}) = 0. \tag{5.4.19}$$

Because $\mathbf{u}^k$, $S^k$ and $p^k$ are independent of $\varepsilon$, by straightforward integration we find that

$$\|\mathbf{v}^{\varepsilon}\|_{\varepsilon} \leqslant C/\varepsilon, \quad \|T^{\varepsilon}\| \leqslant C, \quad |q^{\varepsilon}|_{\varepsilon} \leqslant C. \tag{5.4.20}$$

Let $\varphi \in \mathscr{D}(\Omega)$; then, taking the duality products of (5.1.12) and (5.4.18) with $\varphi \mathbf{v}^{\varepsilon}$ and with $\varepsilon^{-2}\varphi \mathbf{u}^{\varepsilon}$, respectively, by subtraction we get

$$\nu\left(\mathbf{v}^{\varepsilon}\frac{\partial \mathbf{u}^{\varepsilon}}{\partial x_i} - \mathbf{u}^{\varepsilon}\frac{\partial \mathbf{v}^{\varepsilon}}{\partial x_i}, \frac{\partial \varphi}{\partial x_i}\right) + ((\mathbf{u}^{\varepsilon}\nabla)\mathbf{u}^{\varepsilon}, \varphi \mathbf{v}^{\varepsilon}) +$$

$$+ \frac{1}{\varepsilon}(q^{\varepsilon}, \operatorname{div}(\varphi \mathbf{u}^{\varepsilon})) = -\left(\frac{1}{\varepsilon^2}\mathbf{u}^{\varepsilon}, \varphi \mathbf{e}^k\right) + (\tilde{p}^{\varepsilon}, \mathbf{v}^{\varepsilon}\nabla\varphi) +$$

$$+ (1 - \alpha(S^{\varepsilon} + \tau_h), \varphi \mathbf{g}\cdot \mathbf{v}^{\varepsilon}). \tag{5.4.21}$$

With the estimations (5.2.5)–(5.2.8), (5.2.12) and (5.4.20), we have

$$\left|\left(\mathbf{u}^{\varepsilon}\frac{\partial \mathbf{v}^{\varepsilon}}{\partial x_i}-\mathbf{v}^{\varepsilon}\frac{\partial \mathbf{u}^{\varepsilon}}{\partial x_i},\frac{\partial \varphi}{\partial x_i}\right)\right| \leqslant C(|\mathbf{v}^{\varepsilon}|_{\varepsilon}\|\mathbf{u}^{\varepsilon}\|_{\varepsilon}+|\mathbf{u}^{\varepsilon}|_{\varepsilon}\|\mathbf{v}^{\varepsilon}\|_{\varepsilon}) \leqslant C\varepsilon \to 0,$$

$$|((\mathbf{u}^{\varepsilon}\,\nabla)\,\mathbf{u}^{\varepsilon},\varphi\mathbf{v}^{\varepsilon})| \leqslant C\|\mathbf{u}^{\varepsilon}\|_{\varepsilon}^{2}\,\|\varphi\mathbf{v}^{\varepsilon}\|_{\varepsilon} \leqslant C\varepsilon \to 0,$$

$$\left|\frac{1}{\varepsilon}(q^{\varepsilon},\operatorname{div}(\varphi\mathbf{u}^{\varepsilon}))\right| = \left|\frac{1}{\varepsilon}\int_{\Omega_{\varepsilon f}} q^{\varepsilon}\mathbf{u}^{\varepsilon}\nabla\varphi\,\mathrm{d}x\right| \leqslant \frac{C}{\varepsilon}|q^{\varepsilon}|_{\varepsilon}\,|\mathbf{u}^{\varepsilon}|_{\varepsilon} \leqslant C\varepsilon \to 0.$$

Then, passing (5.4.21) to the limit ($\varepsilon \to 0$), we obtain

$$(\mathbf{u}\mathbf{e}^{k},\varphi) = (p,m(\mathbf{u}^{k})\,\nabla\varphi) + (1-\alpha\,(S+\tau_h),\varphi\mathbf{g}\cdot m\,(\mathbf{u}^{k}))$$

because $\mathbf{v}^{\varepsilon} \rightharpoonup m(\mathbf{u}^k)$ weakly in $\mathbf{L}^2(\Omega)$, and hence $(\mathbf{u}, S, p)$ satisfy (5.4.10) in the sense of distributions.

On the other hand, since the vector

$$\sigma_i^{\varepsilon} = a_{ij}\left(\frac{1}{\varepsilon}(\cdot)\right)\frac{\partial}{\partial x_j}(S^{\varepsilon}+\tau_h)$$

is bounded in $\mathbf{L}^2(\Omega)$, it follows that there exist $\sigma_i \in L^2(\Omega)$ for which passing, just in case, to a subsequence, we have $\sigma_i^{\varepsilon'} \rightharpoonup \sigma_i$ weakly in $L^2(\Omega)$ ($i = 1,2,3$). As (5.1.22) can be rewritten in the form

$$(\boldsymbol{\sigma}^{\varepsilon},\nabla T) + (\mathbf{u}^{\varepsilon}/\varepsilon^2,\,T\nabla(S^{\varepsilon}+\tau_h)) = 0, \quad \forall\, T \in H_0^1(\Omega) \tag{5.4.22}$$

and, using convergences already known, we get

$$(\boldsymbol{\sigma},\nabla T) + (\mathbf{u},T\nabla(S+\tau_h)) = 0, \quad \forall\, T \in H_0^1(\Omega). \tag{5.4.23}$$

Taking the duality product of (5.4.19) with $\varphi\,(S^{\varepsilon}+\tau_h)$, we have

$$(a\nabla(S^{\varepsilon}+\tau_h),\,\varphi\,\nabla T^{\varepsilon}) + (\mathbf{a}\,(S^{\varepsilon}+\tau_h)\,\nabla\varphi,\nabla T^{\varepsilon}) = 0. \tag{5.4.24}$$

Also, taking in (5.4.22) $T = \varphi T^{\varepsilon}$ and subtracting this from (5.4.23), we have

$$(\boldsymbol{\sigma}^{\varepsilon},T^{\varepsilon}\nabla\varphi) - ({}^{t}\mathbf{a}\,\nabla T^{\varepsilon},(S^{\varepsilon}+\tau_h)\,\nabla\varphi) = -\,(\mathbf{u}^{\varepsilon}/\varepsilon^2,\varphi\,T^{\varepsilon}\nabla\,(S^{\varepsilon}+\tau_h)) =$$

$$= \int_{\Omega}\frac{1}{\varepsilon^2}\mathbf{u}^{\varepsilon}\,(S^{\varepsilon}+\tau_h)\,\nabla\,(\varphi x_k)\,\mathrm{d}x + \int_{\Omega}\frac{1}{\varepsilon}\mathbf{u}^{\varepsilon}S^{k}\left(\frac{1}{\varepsilon}x\right)\nabla\,(S^{\varepsilon}+\tau_h)\,\mathrm{d}x. \tag{5.4.25}$$

As $({}^{t}\mathbf{a}\,\nabla T^{\varepsilon}) \rightharpoonup m\,({}^{t}\mathbf{a}\,\nabla(S^k+v_k))$ weakly in $\mathbf{L}^2(\Omega)$, and $T^{\varepsilon} \to x_k$ strongly in $L^2(\Omega)$, it follows that

$$(\boldsymbol{\sigma}^{\varepsilon},\,T^{\varepsilon}\nabla\varphi) - ({}^{t}\mathbf{a}\,\nabla T^{\varepsilon},(S^{\varepsilon}+\tau_h)\,\nabla\varphi) \to (\boldsymbol{\sigma},\,x_k\nabla\varphi) +$$

$$+\,(m\,({}^{t}\mathbf{a}\,\nabla(S^k+y_k))\,\nabla\,(S+\tau_h),\,\varphi).$$

From (5.2.12) and (5.2.8) we find that $\mathbf{u}^{\varepsilon}/\varepsilon \to \mathbf{0}$ strongly in $\mathbf{L}^4(\Omega)$, and $\{|\nabla(S^{\varepsilon}+\tau_h)|\}$ bounded; also, $S^k(\varepsilon^{-1}\,(\cdot)) \rightharpoonup m(S^k)$ weakly in $L^4(\Omega)$, and therefore $(\mathbf{u}^{\varepsilon}/\varepsilon^2,\ \varepsilon S^{k}\,(\varepsilon^{-1}(\cdot))\,\nabla(S^{\varepsilon}+\tau_h)) \to 0$; thus (5.4.25) at the limit becomes

$$(\boldsymbol{\sigma},x_k\nabla\varphi) + (m\,({}^{t}\mathbf{a}\,\nabla\,(S^k+y_k))\,\nabla\,(S+\tau_h),\,\varphi) = (\mathbf{u},(S+\tau_h)\,\nabla\,(\varphi x_k)). \tag{5.4.26}$$

Finally, subtracting from (5.4.26) the relation (5.4.23) in which $T = \varphi x_k$, we obtain

$$(\sigma_k, \varphi) = (A_{ki}\frac{\partial}{\partial x_i}(S + \tau_h), \varphi), \quad \forall \varphi \in \mathscr{D}(\Omega).$$

Hence $\sigma_k = A_{ki}\frac{\partial}{\partial x_i}(S + \tau_h)$; that is, calling (5.4.23), $(\mathbf{u}, S, p)$ satisfy (5.4.11) also. □

We have to remark that, besides (5.4.1), the second local-periodical problem which is obtained with the 'heuristic' (see Section 3) two-scale method is not exactly (5.4.2), but the following:

PROBLEM 5.4.2. Find $\chi^l \in H_Y$ $(l = 1,2,3)$ such that

$$\int_Y a_{ij}\frac{\partial \chi^l}{\partial y_j}\frac{\partial T}{\partial y_i}\,\mathrm{d}y = -\int_Y \frac{\partial T}{\partial y_j} a_{jl}\,\mathrm{d}y, \quad \forall T \in H_Y. \tag{5.4.27}$$

Consequently, the definition corresponding to (3.21) of the tensor $\mathbf{a}^h$ is

$$a_{ij}^h = m\left(a_{ij} + a_{ik}\frac{\partial \chi^j}{\partial y_k}\right).$$

We can see that $A_{ij} = a_{ij}^h$ by setting in (5.4.2) $T = \chi^l$, and in (5.4.27) $T = S_k$. This means that the homogenized coefficients $\mathbf{A}$ and $\mathbf{K}$, defined by Theorem 5.4.1, already have the properties proved in Sections 2 and 3: they are positively defined tensors, and in addition $\mathbf{K}$ is symmetric.

Moreover, from the definition of $\mathbf{A}$ we get

$$A_{ij} = \frac{1}{|Y|}\int_Y a_{kl}\frac{\partial(S^i + y_i)}{\partial y_k}\frac{\partial y_j}{\partial y_l}\,\mathrm{d}y. \tag{5.4.28}$$

Notice that (5.4.2) can be put in the form

$$\int_Y a_{ki}\frac{\partial(S^i + y_i)}{\partial y_k}\frac{\partial T}{\partial y_l}\,\mathrm{d}y = 0, \quad \forall T \in H_Y. \tag{5.4.29}$$

Choosing $T = |Y|^{-1}S^j$ in (5.4.29), and adding it to (5.4.28), we obtain

$$A_{ij} = \frac{1}{|Y|}\int_Y a_{kl}\frac{\partial(S^i + y_i)}{\partial y_k}\frac{\partial(S^j + y_j)}{\partial y_l}\,\mathrm{d}y. \tag{5.4.30}$$

Hence, if $\mathbf{a}^\varepsilon$ is symmetric, then $\mathbf{A}$ is also symmetric.

## 6. Boundary Conditions

The convergence of the homogenization method gives us the answer to the following question: what is the condition that must be imposed on an impervious boundary? We have seen that this condition is the slip condition; i.e., the normal component of the velocity must vanish on an impervious boundary.

But, in flow through porous media, there are other types of boundary. In order to obtain such conditions it is necessary to study directly the flow in the vicinity of the boundary.

We start with the boundary between two different porous media. If we take a flat domain $D$ (Fig. 4) around our surface of discontinuity $\Gamma$, with thickness of the order of $2\varepsilon$ and with the other dimensions of the order of unity, we are in the case of 'local periodicity'. Consequently we can apply the reasoning used in Section 2 to find expressions for the conservation of mass and momentum. If we denote by $[\cdot]$ the jump across the surface $\Gamma$, we have

$$\int_{\Gamma} [\mathbf{u} \cdot \mathbf{n}] \, \mathrm{d}s = 0. \tag{6.1}$$

For the conservation of the momentum it is possible to prove that only the pressure is of the order of unity, the other terms being negligibly small. Then we can write

$$\int_{\Gamma} [p^0] \, \mathrm{d}s = 0 \, . \tag{6.2}$$

From (6.1) and (6.2) we obtain the classical conditions

$$[\mathbf{u} \cdot \mathbf{n}] = 0, \qquad [p^0] = 0, \tag{6.3}$$

i.e., the continuity of the normal components of the velocity and the continuity of the pressure.

The second problem is the problem of the interface conditions to be imposed at the boundary between a fluid in a porous medium and a free fluid. In this case we distinguish two situations: (1) the free fluid is in motion which it imparts to the fluid in the porous medium (a typical example is an injection well); (2) the motion of the fluid in porous media induces a motion in the free fluid (two

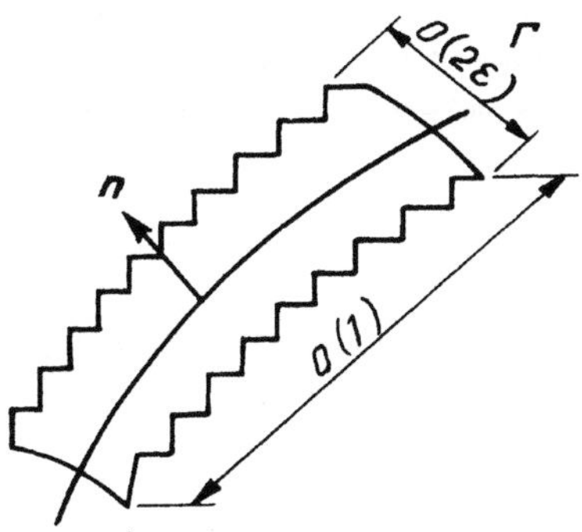


Fig. 4. Representation of boundary between two porous media.

examples are provided by a production well, and a cavity in a porous medium).

We denote by $B$ the porous domain, and by $D$ the free fluid.

The formal expansions (2.6) and (2.7) are appropriate for region $B$. In case 1, for region $D$ we have expansions of the form

$$v_i^\varepsilon(x) = v_i^D(x) + \cdots, \qquad p^\varepsilon(x) = p^D(x) + \cdots, \tag{6.4}$$

which are independent of ε. The expansions (6.4) must match on $S$, the interface between $B$ and $D$, with the expansions (2.6) and (2.7). The matching of the velocity vector implies that

$$\mathbf{v}^D|_S = 0. \tag{6.5}$$

This is the boundary condition to be imposed on $S$ if $B$ is an impervious body. This is natural because the velocity inside $B$ is much smaller than $\mathbf{v}^D$. The matching of $p^\varepsilon$ shows that the appropriate boundary condition for $p^\varepsilon$ is the Dirichlet condition

$$p^0|_S = p^D|_S, \tag{6.6}$$

where the right-hand side is known. Then, for case 1 the appropriate boundary condition for an interface is that the pressure must assume a particular value. (This is so for the injection well.)

In case 2 an expansion of the type (6.4) is impossible. For $B$ (2.6) and (2.7) apply, while for $D$

$$v_i^\varepsilon(x) = \varepsilon^n v_i^D(x) + \cdots, \qquad p^\varepsilon(x) = p^D(x) + \cdots . \tag{6.7}$$

Note that $v^D$ and $p^D$ do not depend on the variable $y$, and consequently the expansions (6.7), (2.6) and (2.7) are of a very different nature. If we introduce (6.7) into the first-order Stokes equation, we have

$$0 = -\operatorname{grad} p^D + \rho\,\mathbf{f}. \tag{6.8}$$

This shows that $p^D$ is in fact the hydrostatic pressure associated with zero velocity, i.e. the free fluid is at rest.

Of course, the problem of the interface between a free fluid and a fluid in a porous medium may be regarded as a limiting case of the conditions (6.3). The conclusion is the same, but the differences between cases 1 and 2 make this difficult to demonstrate directly.

## 7. Dimensionless Parameters

In vectorial form the complete system of equations obtained in Sections 2 and 3 is of the form

$$\mathbf{v} = -\frac{k}{\mu}(\operatorname{grad} p - \rho\,\mathbf{f}). \tag{7.1}$$

$$\operatorname{div}(\rho\mathbf{v}) = 0, \tag{7.2}$$

$$\rho = \rho_0 (1 - \alpha T + \beta p), \tag{7.3}$$

$$p c_f \, \mathbf{v} \operatorname{grad} T = \frac{\mu}{k} \mathbf{v}^2 + \chi^h \Delta T, \tag{7.4}$$

in the steady case, and

$$\mathbf{v} = - \frac{k}{\mu} (\operatorname{grad} p - \rho \, \mathbf{f}), \tag{7.5}$$

$$\frac{\partial (n\rho)}{\partial t} + \operatorname{div} (\rho \mathbf{v}) = 0, \tag{7.6}$$

$$\rho = \rho_0 (1 - \alpha T + \beta p), \tag{7.7}$$

$$[n \rho_f c_f + (1 - n) \rho_s c_s] \frac{\partial T}{\partial t} + \rho_f c_f \mathbf{v} \operatorname{grad} T = \frac{\mu}{k} \mathbf{v}^2 + \chi^h \Delta T, \tag{7.8}$$

in the non-steady case.

Either system (7.1)−(7.4) or (7.5)−(7.8) may be used in the study of underground combustion.

In the case of natural convection in porous media, the system of equations obtained in Section 4 is in vectorial form,

$$\mathbf{v} = \frac{k}{\mu} (\operatorname{grad} p + \alpha \rho T \mathbf{g}), \tag{7.9}$$

$$\operatorname{div} \mathbf{v} = 0, \tag{7.10}$$

$$\rho = \rho_0 (1 - \alpha T), \tag{7.11}$$

$$\rho c_f \mathbf{v} \operatorname{grad} T = \chi^h \Delta T, \tag{7.12}$$

in the steady case; in the non-steady case there is in addition the modified energy equation

$$[n \rho_f c_f + (1 - n) \rho_s c_s] \frac{\partial T}{\partial t} + \rho_f c_f \mathbf{v} \operatorname{grad} T = \chi^h \Delta T. \tag{7.13}$$

In order to establish the physical significance of the various terms in these systems of equations, we take a characteristic pore length $l$, a characteristic length $L$ of the domain $\Omega$ and a characteristic velocity $Q$. Now, the small parameter is well defined $\varepsilon = l/L$, and it is known (Bear (1972), Ene and Sanchez-Palencia (1975), Ene and Gogonea (1973)) that the permeability is of the from $k = k^* l^2$, where $k^*$ is a dimensionless permeability. It is possible then to introduce the following dimensionless numbers:

$$R_\varepsilon = \frac{Q \rho_0 l}{\mu}, \quad \text{the Reynolds number,} \tag{7.14}$$

$$P = \frac{\mu c}{\chi}, \quad \text{the Prandtl number}, \tag{7.15}$$

$$S_\varepsilon = \frac{\mu Q^2}{\chi(T - T_r)}, \tag{7.16}$$

$$G_\varepsilon = \frac{g\alpha\rho_0^2 l^3 (T - T_r)}{\mu^2}, \text{ the Grashof number,} \tag{7.17}$$

$$Ra_\varepsilon = G_\varepsilon \cdot P, \quad \text{the Rayleigh number,} \tag{7.18}$$

where $T_r$ is a reference temperature.
Then, Darcy's law holds for $R_\varepsilon \sim O(\varepsilon^{-1})$ or, in an equivalent form,

$$\frac{l^2 \rho_0 Q}{\mu L} \sim O(1). \tag{7.19}$$

Equation (7.4) makes sense for $S_\varepsilon \sim O(\varepsilon^2)$ and $P \sim O(\varepsilon^2)$, or

$$\frac{\mu L^2 Q^2}{l^2 \chi (T - T_r)} \sim O(1), \tag{7.20}$$

$$\frac{\mu c L^2}{\chi l^2} \sim O(1). \tag{7.21}$$

If we take into consideration the range of validity of Darcy's law, the conditions (7.20) and (7.21) may be combined in the form

$$PR_\varepsilon \sim \varepsilon^{-1} S_\varepsilon. \tag{7.22}$$

In the case of natural convection in porous media, Equation (7.9) gives us $P \cdot R_\varepsilon \sim O(\varepsilon)$, and from (7.12) then $Ra_\varepsilon \sim O(\varepsilon)$, or

$$\frac{Q \rho_0 L c}{\chi} \sim O(1), \tag{7.23}$$

$$\frac{\rho_0^2 g \alpha l^2 L c (T - T_r)}{\mu \chi} \sim O(1). \tag{7.24}$$

The Rayleigh number (7.18) is defined at the microscopic level (with $l^3$). In the study of natural convection in porous media it is usual to define the Rayleigh number with two scales ($l^2 L$ instead of $l^3$):

$$Ra = \frac{\rho_0 \alpha g l^2 L (T - T_r)}{\mu \lambda}, \quad \lambda = \frac{\chi}{\rho c}. \tag{7.25}$$

From (7.25) and (7.24) it is clear that the previous condition may be written in the form $Ra \sim O(1)$; this is in good agreement with experimental results (see Chapter 3).

## 8. Comments and Bibliographical Notes

Extensive mathematical studies of the homogenization method are to be found in Bensoussan *et al.* (1978) and Sanchez-Palencia (1980). Sanchez-Palencia also gives numerous examples of application of this method in fields such as mechanics, acoustics and electromagnetism. Section 1.1, which is a classical treatment, is based on Chap. 5 of Sanchez-Palencia.

Section 1.2 is based on Ene and Sanchez-Palencia (1973, 1975). The problem is also dealt with by Ene (1978, 1984), Lions (1978) and Sanchez-Palencia (1980, Chap. 7). The results given in Sections 1.3 and 1.4 are from Ene and Sanchez-Palencia (1981 a, b, c, 1982) and Ene (1984).

The convergence of the homogenization process was proved for the bidimensional case by Tartar (1980), and for the real case – Section 1.5 – by Poliševski (1984, 1986b).

The determination of boundary conditions on different interfaces is a classical problem (Bear (1972), Ene and Gogonea (1973), Gheorghiţă (1966), Muskat (1946)). The discussion in Section 1.6 shows that the case of interface between a porous body and a free fluid is a difficult problem, and several situations may appear following different orders of magnitude. For other studies of this problem, see Ene and Sanchez-Palencia (1975), Levy and Sanchez-Palencia (1975), and Sanchez-Palencia (1980).

The dimensionless parameters are treated in a classical manner. The order of magnitude for the Darcy's law appears in Ene and Sanchez-Palencia (1975) and Levy and Sanchez-Palencia (1975). For thermal problems in porous media they appear in this form in Ene and Sanchez-Palencia (1982).

Of course, the problem of formulating the system of equations governing flow through porous media and boundary conditions may be treated in classical manner. Our presentation is different because we use a new method, the homogenization method, which enables us to obtain a rigorous mathematical description of the problem.

CHAPTER 2

# Natural Convection in Bounded Domains

In a rigid, porous medium, gravitational forces and density differences due to temperature gradients can set a viscous fluid in motion. Attempts to describe this motion have always met serious difficulties. Supposing that the skeleton is fixed, we can accept that the velocity of the fluid is far lower than the acoustic velocity, and thus that the motion has little effect on the pressure; therefore we neglect variations of the thermodynamic quantities resulting from pressure changes. Moreover, we assume that the temperature differences are small enough to use the Boussinesq approximation; that the density of the gravitational force varies affinely with the temperature. We adopt the model proposed in Chapter 1, which was obtained by a homogenization process for which we have proved the convergence; we refer to it sometimes as the Darcy-Boussinesq system.

## 1. The Steady Case. Basic Results

This section deals with basic aspects of the study of steady convection in fluid-saturated porous media. We prove the existence of the weak solutions, via a theorem of the Gossez type (Gossez (1966)), and give some regularity properties including a natural maximum principle.

Let $\Omega$ be an open, connected bounded set in $\mathbb{R}^n$ ($n = 2, 3$) locally located on one side of the boundary $\partial\Omega$ – a manifold of class $C^2$, composed of a finite number of connected components. The Darcy's velocity $\mathbf{u}$, the pressure $p$ and the temperature $T$ must satisfy in some way the system

$$\operatorname{div} \mathbf{u} = 0 \quad \text{in } \Omega , \tag{1.1}$$

$$\mathbf{Bu} + \nabla p = (1 - \alpha (T - T_r))\, \mathbf{g} \quad \text{in } \Omega , \tag{1.2}$$

$$- \operatorname{div} (\mathbf{A} \nabla T) + \mathbf{u} \nabla T = 0 \quad \text{in } \Omega , \tag{1.3}$$

with the boundary conditions

$$\mathbf{u} \cdot \mathbf{v} = 0 \quad \text{on } \partial\Omega , \tag{1.4}$$

$$T = \tau \quad \text{on } \partial\Omega , \tag{1.5}$$

where $\mathbf{B}$ ($\mathbf{K} = \mathbf{B}^{-1}$) is the positive symmetric constant tensor of permeability, $\alpha > 0$ is the volumetric coefficient of thermal expansion, $\mathbf{g} \in H^2(\Omega)$ is the potential type gravitational acceleration, **A** is the positive constant tensor of thermal diffusion, $\boldsymbol{\nu}$ is the unit outward normal on $\partial\Omega$, $\tau \in H^{3/2}(\partial\Omega)$ is the non-uniform temperature of the boundary (if $\tau$ is uniform the problem is trivial) and $T_r > 0$ is a uniform reference temperature; for convenience

$$T_r = \frac{1}{2}(\sup_{x\in\partial\Omega}\tau + \inf_{x\in\partial\Omega}\tau).$$

As usual, the scalar products and norms in $L^2(\Omega)$, $H^m(\Omega)$ and $H_0^1(\Omega)$ are denoted by

$$(u, v) = \int_\Omega u\cdot v\,\mathrm{d}x, \quad |u| = (u,u)^{1/2},$$

$$((u, v))_m = \sum_{|j|\leqslant m} (D^j u, D^j v), \qquad \|u\|_m = ((u,u))_m^{1/2},$$

$$((u,v)) = (\nabla u, \nabla v), \quad \|u\| = ((u,u))^{1/2},$$

and the norm in $L^p(\Omega)$ $(p\neq 2)$ by $|\cdot|_p$. We use the same notations for the scalar products and norms in $\mathbf{L}^2(\Omega) = [L^2(\Omega)]^n$, $\mathbf{H}^m(\Omega)$, $\mathbf{H}_0^1(\Omega)$ and $\mathbf{L}^p(\Omega)$.

Now, in order to derive the homogeneous boundary conditions, using Lemma 5.1.1 (Chapter 1), we introduce for any $h > 0$ an element $\tau_h \in H^2(\Omega)$ with the properties

$$\tau_h = \tau - T_0 \quad \text{on } \partial\Omega, \tag{1.6}$$

$$|S\nabla\tau_h| \leqslant h\|S\|, \quad \forall S \in H_0^1(\Omega). \tag{1.7}$$

Setting $s = T - (\tau_h + T_r)$ (later we shall choose a suitable value for the parameter $h > 0$), the system (1.1)–(1.5) becomes

$$\operatorname{div}\mathbf{u} = 0 \quad \text{in } \Omega, \tag{1.8}$$

$$\mathbf{B}\mathbf{u} + \nabla p + \alpha(s + \tau_h)\mathbf{g} = 0 \quad \text{in } \Omega, \tag{1.9}$$

$$-\operatorname{div}(\mathbf{A}\nabla(s+\tau_h)) + \mathbf{u}\nabla(s+\tau_h) = 0 \quad \text{in } \Omega, \tag{1.10}$$

$$\mathbf{u}\cdot\mathbf{v} = 0 \quad \text{on } \partial\Omega, \tag{1.11}$$

$$s = 0 \quad \text{on } \partial\Omega. \tag{1.12}$$

For practical reasons we write the system (1.8)–(1.12) in dimensionless form, first defining

$$\mathbf{x}^* = L^{-1}\mathbf{x}, \quad \mathbf{x}^* \in \Omega^* \Leftrightarrow \mathbf{x} \in \Omega,$$

$$s^* = \beta^{-1}s, \qquad \tau_h^* = \beta^{-1}\tau_h, \qquad T_r^* = \beta^{-1}T_r,$$

$$\mathbf{u}^* = La_1^{-1}\mathbf{u}, \qquad p^* = a_1^{-1}b_1^{-1}p,$$

$$\mathbf{B}^* = b_1^{-1}\mathbf{B}, \qquad \mathbf{A}^* = a_1^{-1}\mathbf{A}, \qquad \mathbf{g}^* = g^{-1}\mathbf{g},$$

where $L$ is the length of the edge of the $n$-cube in which $\Omega$ can be included,

$$\beta = \sup_{x\epsilon\partial\Omega}\tau - \inf_{x\epsilon\partial\Omega}\tau ,$$

$g = |\mathbf{g}|_\infty$, and $a_1$ and $b_1$ are respectively the smallest eigenvalues of **A** and **B**. If we define the dimensionless Rayleigh number by

$$a^* = a_1^{-1}b_1^{-1}gL\alpha\beta ,$$

the system (1.8)–(1.12) takes the form

$$\operatorname{div} \mathbf{u} = 0 \quad \text{in } \Omega , \tag{1.13}$$

$$\mathbf{Bu} + \nabla p + a(s + \tau_h)\,\mathbf{g} = 0 \quad \text{in } \Omega , \tag{1.14}$$

$$- \operatorname{div} (\mathbf{A}\nabla(s + \tau_h)) + \mathbf{u}\nabla(s + \tau_h) = 0 \quad \text{in } \Omega , \tag{1.15}$$

$$\mathbf{u}\cdot\mathbf{v} = 0 \quad \text{on } \partial\Omega , \tag{1.16}$$

$$s = 0 \quad \text{on } \partial\Omega . \tag{1.17}$$

Here, and from now on, we omit the asterisks to simplify the notation.

REMARK 1.1.

$$\sup_{x\epsilon\partial\Omega}\tau_h(x) = \frac{1}{2}, \qquad \inf_{x\epsilon\partial\Omega}\tau_h(x) = -\frac{1}{2} .$$

The proof of the existence of the weak solutions of the problem (1.13)–(1.17) is based on the following theorem, a slight generalization of the theorem of Gossez (1966).

THEOREM 1.1 *Let $Y'$ be the dual space of $Y$, which is a separated locally convex space, continuously embedded in the reflexive Banach space $X$. If the mapping $G: X \to Y'$ is weakly continuous (that is, continuous between the weak topologies) and coercive (that is, $\exists\, r > 0$ such that*

$$< Gy, y > \geqslant 0, \qquad \forall y \epsilon Y, \qquad |y| = r,$$

*where $|\cdot|$ is the norm in $X$), then $\exists\, x_0 \epsilon \dot{B}_r = \{x\epsilon X|\, |x| \leqslant r\}$ such that $Gx_0 = 0$.*

*Proof.* For every $F \epsilon \mathscr{F}$, where

$$\mathscr{F} = \{E\,|\,E \text{ finite-dimensional subspace of } Y\},$$

$j_F: F \to Y$ is the natural injection, $j_F': Y' \to F'$ the surjection defined by

$$< j_F' y', y > = < y', y >, \quad \forall y' \epsilon Y', \forall y \epsilon F,$$

and $G_F: F \to F'$ the continuous operator given by $G_F = j_F' \circ G \circ j_F$.

Suppose that $G_F y \neq 0$, $\forall y \in F \cap B_r$. As $F$ is closed in $Y$, and the topology induced on $F$ is Euclidean, the set $F \cap B_r$ is compact in $F$. Identifying $F'$ with $F$, we define $T: F \cap B_r \to F \cap B_r$ by $T_y = -r\,|G_F y|^{-1} G_F y$. As $T$ is continuous and $F \cap B_r$ is convex, by Brouwer's fixed-point theorem, there exists $y_0 \in F \cap B_r$, such that $Ty_0 = y_0$. It follows that

$$< Gy_0, y_0 > \; = \; < G \circ j_F(y_0), y_0 > \; = \; < G_F y_0, y_0 > =$$
$$= -r^{-1} |G_F y_0| < Ty_0, y_0 > \; = \; -r\,|G_F y_0| < 0 ,$$

which is in contradiction with the hypothesis, since $y_0 \in Y$ and $|y_0| = |Ty_0| = r$. Hence, for any $F \in \mathcal{F}$ there exists $y_F \in F \cap B_r$ such that $G_F y_F = 0$. (It is obvious that if $Y$ is finite-dimensional, then the theorem is already proved.)

Now we can define for every $E \in \mathcal{F}$ the non-empty set

$$V_E = \bigcup_{F \in \mathcal{F} | F \supseteq E} \{ y_F \in F \cap B_r \mid G_F y_F = 0 \} ,$$

$\tilde{V}_E$ being the weak closure of $V_E$ in $X$. As $B_r$ is weakly compact in $X$ and the family $\{ \tilde{V}_E \}_{E \in \mathcal{F}} \subseteq B_r$ has the property that every finite intersection of its sets is non-empty, it follows that $\exists\, x_0 \in \bigcap_{E \in \mathcal{F}} \tilde{V}_E$.

For any $y \in Y$ we choose $E(y) \in \mathcal{F}$ with the property $y \in E(y)$. As $x_0 \in \tilde{V}_{E(y)}$ there exists a (generalized) sequence $\{y_\delta\}_{\delta \in \Delta}$ (which depends on $y$), included in $V_{E(y)}$ and weakly converging to $x_0$. By using the definition of $V_{E(y)}$ it can be shown that for any $\delta \in \Delta$ there exists $F_\delta \in \mathcal{F}$ with the properties $F_\delta \supseteq E(y)$ and $G_{F_\delta} y_\delta = 0$. Consequently,

$$< Gy_\delta, y > \; = \; < G \circ j_{F_\delta}(y_\delta), j_{F_\delta}(y) > \; = \; < G_{F_\delta} y_\delta, y > \; = 0 ,$$

and because $G$ is weakly continuous we can pass to the limit in the last relation; thus

$$< Gx_0, y > \; = 0, \quad \forall y \in Y. \qquad \square$$

We now make the correspondence between the elements of the problem (1.13)–(1.17) and those of Theorem 1.1. We introduce the space

$$\mathbf{H} = \{ \mathbf{v} \in \mathbf{L}^2(\Omega) \mid \operatorname{div} \mathbf{v} = 0 \text{ in } \Omega, \; \mathbf{v} \cdot \boldsymbol{\nu} = 0 \text{ on } \partial\Omega \} .$$

Then, instead of $X$, we have the Hilbert space $\mathbf{H} \times H_0^1(\Omega)$ with the scalar product

$$((\mathbf{u}, s), (\mathbf{v}, t))_x = (\mathbf{u}, \mathbf{v}) + ((s, t)),$$

and instead of $Y$ we have the Banach space $\mathbf{H} \times \mathring{W}_4^{(1)}(\Omega)$ with the usual norm on a product space; the operator $G$ is defined here for any $(\mathbf{u}, s) \in \mathbf{H} \times H_0^1(\Omega)$ and for any $(\mathbf{v}, t) \in \mathbf{H} \times \mathring{W}_4^{(1)}(\Omega)$ by

$$< G(\mathbf{u}, s), (\mathbf{v}, t) > \; = a^2 A(s + \tau_h, t) + B(\mathbf{u}, \mathbf{v}) + a(s + \tau_h, \mathbf{g} \cdot \mathbf{v}) +$$
$$+ . a^2 b(\mathbf{u}, s + \tau_h, t) , \tag{1.18}$$

where $A(s,t) = (\mathbf{A}\nabla s, \nabla t)$, $B(\mathbf{u},\mathbf{v}) = (\mathbf{Bu},\mathbf{v})$ and $b(\mathbf{u}, s,t) = (\mathbf{u}, t\nabla s)$.

We have arrived at the weak formulation of the problem (1.13)−(1.17): to find $(\mathbf{u}, s)\in\mathbf{H} \times H_0^1(\Omega)$ such that $\forall (\mathbf{v}, t)\in\mathbf{H} \times \mathring{W}_4^{(1)}(\Omega)$ the condition

$$< G(\mathbf{u},s),(\mathbf{v},t) > = 0 \tag{1.19}$$

holds.

The 'equivalence' between the problems (1.13)−(1.17) and (1.19) is demonstrated by the following two propositions.

PROPOSITION 1.1. *If* $(\mathbf{u},s,p)$ *are smooth functions satisfying* (1.13)−(1.17), *then from the dual product of* (1.14) *and* (1.15) *with* $\mathbf{v}\in\mathbf{H}$ *and* $t\in \mathring{W}_4^{(1)}(\Omega)$ *it obviously follows that* $(\mathbf{u}, s)$ *is a solution of* (1.19).

PROPOSITION 1.2. *If* $(\mathbf{u}, s)\in \mathbf{H} \times H_1^0(\Omega)$ *satisfies* (1.19), *then, provided functions are chosen in a proper manner, we get*

$$B(\mathbf{u},\mathbf{v}) + a(s + \tau_h, \mathbf{g}\cdot\mathbf{v}) = 0, \quad \forall\, \mathbf{v} \in \mathbf{H}, \tag{1.20}$$

$$A(s + \tau_h, t) + b(\mathbf{u}, s + \tau_h, t) = 0 \quad \forall t \in \mathring{W}_4^{(1)}(\Omega). \tag{1.21}$$

As $(\mathbf{Bu} + a(s + \tau_h)\,\mathbf{g}) \in \mathbf{L}^2(\Omega)$ and the orthogonal complement of $\mathbf{H}$ in $\mathbf{L}^2(\Omega)$ is

$$\mathbf{H}^\perp = \{\mathbf{w} \in \mathbf{L}^2(\Omega) \mid \exists p \in H^1(\Omega) \ \text{ such that } \mathbf{w} = \nabla p\}$$

(for related properties the reader is refered to Temam (1977, Chap. 1)), it follows from (1.20) that $\exists\, p\in H^1(\Omega)$ such that (1.14) is satisfied in $\mathbf{L}^2(\Omega)$. Also, because

$$|b(\mathbf{u}, s + \tau_h, t)| \leqslant c_0\, |\mathbf{u}|\, |\nabla(s + \tau_h)|\; |t|_\infty, \tag{1.22}$$

and $\mathring{W}_4^{(1)}(\Omega)\subseteq L^\infty(\Omega)$ it follows from (1.21) that (1.15) is satisfied in $W_{4/3}^{(-1)}(\Omega)$. Finally, from the definitions of $\mathbf{H}$ and $H_0^1(\Omega)$ it follows that $\mathbf{u}$ and $s$ satisfy (1.13) in the distribution sense, and (1.16) and (1.17) in the trace senses of the spaces $\mathbf{H}$ and $H_0^1(\Omega)$, respectively. □

The following theorem is the main result of this section:

THEOREM 1.2. *The problem* (1.19) *has at least one solution.*

*Proof.* We shall check that $G$ defined by (1.18) is

(a) weakly continuous
(b) coercive;

thus the result is a straight consequence of Theorem 1.1.

(*a*) *Let* $(\mathbf{u}_k, s_k) \xrightarrow{(k\to\infty)} (\mathbf{u}, s)$ weakly in $\mathbf{H}\times H_0^1(\Omega)$. Obviously, we have only to prove that

$$b(\mathbf{u}_k, s_k, t) \to b(\mathbf{u}, s, t), \quad \forall t \in \mathring{W}_4^{(1)}(\Omega),$$

all the other convergences being trivial.

First we note that

$$b(\mathbf{u}_k, s_k, t) - b(\mathbf{u}, s, t) = b(\mathbf{u}_k, s_k - s, t) - b(\mathbf{u}_k - \mathbf{u}, t, s).$$

As $(s\nabla t) \in \mathbf{L}^2(\Omega)$, it follows that $b(\mathbf{u}_k - \mathbf{u}, t, s) \to 0$. For the first term on the right-hand side we have the estimation

$$\| b(\mathbf{u}_k, s_k - s, t) | \leqslant |\mathbf{u}_k| \, |\nabla t|_4 | s_k - s|_4 \leqslant C_1 |\mathbf{u}_k| \, |t|_{W_4^1(\Omega)} | s_k - s|_4. \tag{1.23}$$

Since $\{|\mathbf{u}_k|\}_k$ is bounded as $\{\mathbf{u}_k\}_k$ is weakly convergent in $\mathbf{H}$, and $|s_k - s|_4 \to 0$ as the embedding of $H_0^1(\Omega)$ in $L^4(\Omega)$ is compact, from (1.23) it follows that $b(\mathbf{u}_k, s_k - s, t) \to 0$.

(b) From (1.18), for any $(\mathbf{u}, s) \in \mathbf{H} \times \mathring{W}_4^{(1)}(\Omega)$ we have

$$< G(\mathbf{u}, s), (\mathbf{u}, s) > = a^2 A(s,s) + B(\mathbf{u},\mathbf{u}) - a^2 (\operatorname{div}(\mathbf{A}\nabla\tau_h), s) +$$

$$+ a(s, \mathbf{g}\cdot\mathbf{u}) + a(\tau_h, \mathbf{g}\cdot\mathbf{u}) + a^2 b(\mathbf{u}, \tau_h s) \geqslant a^2\|s\|^2 + |\mathbf{u}|^2 - a^2 |\operatorname{div}(\mathbf{A}\nabla\tau_h)| \, |s| -$$

$$- a|s| \, |\mathbf{u}| - a|\tau_h| \, |\mathbf{u}| - a^2 |\mathbf{u}| \, |s\nabla\tau_h| . \tag{1.24}$$

Using the property (1.7) of $\tau_h$ and the Friedrichs inequality

$$|s| \leqslant 2a_0^{-1}\|s\|, \quad \forall s \in H_0^1(\Omega), \ (a_0 \geqslant 2\pi\sqrt{n}), \tag{1.25}$$

the relation (1.24) becomes

$$< G(\mathbf{u}, s), (\mathbf{u}, s) > \geqslant a^2 \|s\|^2 + |\mathbf{u}|^2 - (2aa_0^{-1} + ha^2) |\mathbf{u}| \, \|s\| -$$

$$- 2a^2 a_0^{-1} |\operatorname{div}(\mathbf{A}\nabla\tau_h)| \, \|s\| - a|\tau_h| \, |\mathbf{u}| . \tag{1.26}$$

If $h > 0$ is sufficiently small so that

$$h < 2a^{-1} a_0^{-1}(a_0 - 1), \tag{1.27}$$

we obtain immediately from (1.26) the coercive property of $G$. □

We proceed with the regularity properties of the weak solutions.

LEMMA 1.1. *If* $(\mathbf{u}, s) \in \mathbf{H} \times H_0^1(\Omega)$ *is a solution of the problem* (1.19), *then* $\mathbf{u} \in \mathbf{H}^1(\Omega)$ *and*

$$\|\mathbf{u}\|_1 \leqslant C_0(\Omega), \tag{1.28}$$

*where* $C_0$ *is a constant* (defined by the data).

*Proof.* From Proposition 1.2, $\exists\, p \in H^1(\Omega)$ so that (1.14) is satisfied in $\mathbf{L}^2(\Omega)$. Taking into account (1.13) and (1.16), it follows that $p$ has to satisfy

$$- \operatorname{div}(\mathbf{K}\nabla p) = a \operatorname{div}(\mathbf{K}\mathbf{g}(s + \tau_h)) \quad \text{in } \Omega, \tag{1.29}$$

$$(\mathbf{K}\nabla p)\mathbf{v} = - a(\tau - T_0)(\mathbf{K}\mathbf{g}) \quad \text{on } \partial\Omega . \tag{1.30}$$

The system represented by (1.29) and (1.30) is a Neumann problem; as the compatibility condition is obviously satisfied, there exists a unique $p$ (up to an additive constant) which satisfies (1.29) and (1.30). Moreover, as

$$\operatorname{div}\left(\mathbf{Kg}(s+\tau_h)\right) = (s+\tau_h)\,K_{ij}\frac{\partial g_j}{\partial x_j} + K_{ij}\,g_j\frac{\partial(s+\tau_h)}{\partial x_i} \in L^2(\Omega),$$

$$(\tau - T_0)\,(\mathbf{Kg})\cdot\mathbf{v} \in H^{3/2}(\partial\Omega) \subseteq H^{1/2}(\partial\Omega),$$

by usual regularity results (see for the general case Lions and Magenes (1968), Chap. 2, Sect. 7) for the Neumann problem we get $p \in H^2(\Omega)$ and

$$\|p\|_2 \leqslant c_2\,\|s\| + c_3, \tag{1.31}$$

where the $c_i$ are constants. Now, from (1.14) we have $\mathbf{u} \in \mathbf{H}^1(\Omega)$, and we can replace the inequality (1.22) by

$$|b(\mathbf{u}, s+\tau_h, t)| \leqslant |\mathbf{u}|_4\,|\nabla(s+\tau_h)|\,|t|_4 \leqslant c_4\,\|\mathbf{u}\|_1\,|\nabla(s+\tau_h)|\,\|t\|,$$
$$\forall t \in H_0^1(\Omega). \tag{1.32}$$

Hence (1.15) is actually satisfied in $H^{-1}(\Omega)$, and therefore

$$< G(\mathbf{u}, s), (\mathbf{v}, t) > \; = 0 \quad \forall (\mathbf{v}, t) \in \mathbf{H} \times H_0^1(\Omega). \tag{1.33}$$

Setting $(\mathbf{v}, t) = (\mathbf{u}, s)$ in (1.33), we obtain

$$|\mathbf{u}|^2 + \|s\|^2 \leqslant c_5\,|\mathbf{u}| + c_6\,\|s\|,$$

and hence

$$\|s\| \leqslant c_1(\Omega). \tag{1.34}$$

Then, via (1.31), we find that $p$ is bounded in $H^2(\Omega)$. Thus the relation (1.28) can be obtained directly from (1.14). □

The following weak maximum principle is formulated in terms of inequalities in the sense of $H^1(\Omega)$, as presented in Chapter 1, Section 5.2:

THEOREM 1.3. *If* $(\mathbf{u}, s) \in \mathbf{H} \times H_0^1(\Omega)$ *is a solution of the problem* (1.19), *then* $s \in L^\infty(\Omega)$ *and*

$$|s + \tau_h|_\infty \leqslant \frac{1}{2}. \tag{1.35}$$

*Proof.* Setting in (1.33) $\mathbf{v} = \mathbf{0}$, we get

$$A(s+\tau_h, t) = b(\mathbf{u}, t, s+\tau_h), \quad \forall t \in H_0^1(\Omega). \tag{1.36}$$

Via Remark 1.1, we have from Proposition 5.2.2 in Chapter 1 that $R = \max\{s + \tau_h - \frac{1}{2}, 0\} \in H_0^1(\Omega)$, and from Proposition 5.2.3 in Chapter 1

$$\nabla R = \begin{cases} \nabla(s+\tau_h), & R \neq 0, \\ \mathbf{0} & R = 0. \end{cases}$$

Setting $t = R$ in (1.36), we obtain

$$\|R\|^2 \leqslant A(R,R) = A(s+\tau_h, R) = b(\mathbf{u}, R, s+\tau_h) = b(\mathbf{u}, R, R) = 0,$$

and hence $R=0$ in $H_0^1(\Omega)$; that is $(s+\tau_h) \leqslant \frac{1}{2}$ on $\Omega$ in $H^1(\Omega)$. According to Proposition 5.2.1 in Chapter 1, it follows that

$$s + \tau_h \leqslant \frac{1}{2}, \quad \text{a.e. on } \Omega .$$

Analogously, with $R = \min\{s + \tau_h + \frac{1}{2}, 0\}$, we get

$$s + \tau_h \geqslant -\frac{1}{2}, \quad \text{a.e. on } \Omega .$$

Thus (1.35) is proved, as is the whole theorem, because $\tau_h \in H^2(\Omega) \subseteq L^\infty(\Omega)$. □

LEMMA 1.2. *If* $(\mathbf{u}, s) \in \mathbf{H} \times H_0^1(\Omega)$ *is a solution of the problem* (1.19), *then for any subdomain* $\Omega' \subseteq \bar{\Omega}' \subseteq \Omega$ *we have* $s \in H^2(\Omega')$ *and*

$$\|s\|_{H^2(\Omega')} \leqslant c(\Omega'), \quad c(\Omega') = \text{const.} \tag{1.37}$$

*Proof.* Let $\Omega''$ have the properties $\bar{\Omega}' \subseteq \Omega'' \subseteq \bar{\Omega}'' \subseteq \Omega$ and $t \in L^2(\Omega'')$. Then, for any $\varepsilon \in \mathbb{R}$ and $i$, $1 \leqslant i \leqslant n$, we define

$$\tau_\varepsilon^i t(x) = t(x + \varepsilon e_i)$$

where $e_i$ is the unit vector of the $Ox_i$ axis. Note that

$$(\tau_\varepsilon^i t, R) = (t, \tau_{-\varepsilon}^i R), \quad \forall t, R \in L^2(\Omega'').$$

Finally, we introduce the parameter

$$\Delta_\varepsilon^i t = \frac{1}{\varepsilon}(\tau_\varepsilon t - t).$$

Now, let $\varepsilon$ be such that $|\varepsilon| < d(\bar{\Omega}'', \partial\Omega)$. Choosing successively in (1.19) the test functions $(\mathbf{0}, t)$ and $(\mathbf{0}, \tau_{-\varepsilon}^i t)$ with $t \in H_0^1(\Omega'')$, we obtain

$$A(s,t) + b(\mathbf{u}, s, t) = (\operatorname{div}(\mathbf{A}\nabla\tau_h), t) - b(\mathbf{u}, \tau_h, t), \tag{1.38}$$

$$A(\tau_\varepsilon^i s, t) + b(\tau_\varepsilon^i \mathbf{u}, \tau_\varepsilon^i s, t) = (\operatorname{div}(\mathbf{A}\nabla\tau_h), \tau_{-\varepsilon}^i t) - b(\mathbf{u}, \tau_h, \tau_{-\varepsilon}^i t). \tag{1.39}$$

Subtracting (1.38) from (1.39), and dividing by $\varepsilon$, we have

$$A(\Delta_\varepsilon^i s, t) + b(\tau_\varepsilon^i \mathbf{u}, \tau_\varepsilon^i s, t) + b(\Delta_\varepsilon^i \mathbf{u}, s, t) = (\operatorname{div}(\mathbf{A}\nabla\tau_h), \Delta_{-\varepsilon}^i t)$$
$$- b(\mathbf{u}, \tau_h, \Delta_{-\varepsilon}^i t). \tag{1.40}$$

With $f \in \mathcal{D}(\Omega'')$, $f = 1$ on $\bar{\Omega}'$, we define $F = fs$. Setting in (1.40) $t = \Delta_\varepsilon^i F$ and taking into account (see Nečas (1967), Chap. 4) that

$$|\Delta_{-\varepsilon}^i t|_{L^2(\Omega)} \leqslant c_1 \|t\|,$$

we obtain the estimation

$$\| \Delta_\varepsilon^i F \| \leqslant |\Delta_\varepsilon^i \mathbf{u}| \, |s|_\infty + c_1 \left( |\operatorname{div} (\mathbf{A} \nabla \tau_h)| + |\mathbf{u}|_4 \, |\nabla \tau_h|_4 \right),$$

and finally

$$\| \Delta_\varepsilon^i F \| \leqslant c_2 \left( \| \mathbf{u} \|_1 \, |s|_\infty + |\operatorname{div} (\mathbf{A} \nabla \tau_h)| + \| \mathbf{u} \|_1 \, \| \tau_h \|_2 \right).$$

From (1.28) and (1.35), it follows that the sequence $\{\Delta_\varepsilon^i F\}_\varepsilon$ is bounded in $H^1(\Omega'')$, so we can extract a subsequence (still denoted $\varepsilon$) such that

$$\Delta_\varepsilon^i F \rightharpoonup R \quad \text{weakly in } H^1(\Omega'').$$

However, as $\Delta_\varepsilon^i F$ converges to $\dfrac{\partial s}{\partial x_i}$ in $\mathscr{D}'(\Omega')$, $\dfrac{\partial s}{\partial x_i} = R \in H^1(\Omega')$ for any $i$, $1 \leqslant i \leqslant n$. □

THEOREM 1.4. *If* $(\mathbf{u}, s) \in \mathbf{H} \times H_0^1(\Omega)$ *is a solution of the problem* (1.19), *then* $s \in H^2(\Omega)$ *and*

$$\| s \|_2 \leqslant c_2(\Omega). \tag{1.41}$$

*Moreover,* $\mathbf{u} \in \mathbf{H}^2(\Omega)$ *and*

$$\| \mathbf{u} \|_2 \leqslant c_3(\Omega). \tag{1.42}$$

*Proof.* With Lemma 1.2 we have established the regularity inside the domain, so it remains to prove the regularity only on the boundary. In view of the local charts and the partition of unity we can confine ourselves (see Nečas (1967), Chap. 4), to the case of an $n$-parallelepipedon $Y$ with $\operatorname{supp} s \subseteq Y \cup \Sigma$ (Fig. 5), $\Sigma = \{x \in \partial Y \mid x_n = 0\}$). By the technique used in Lemma 1.2, one can prove that $\dfrac{\partial^2 s}{\partial x_i \partial x_j} \in L^2(Y)$ if at least one of the indices $i$ and $j$ is different from $n$. From

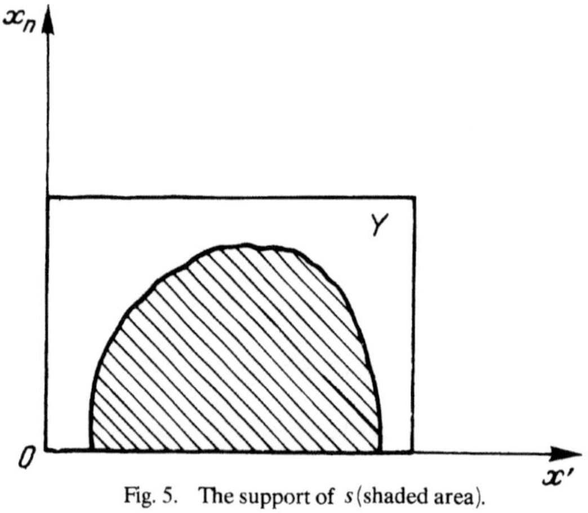


Fig. 5. The support of $s$ (shaded area).

(1.15) it then follows that $\frac{\partial^2 s}{\partial x_n^2} \in L^{3/2}(Y)$, and hence $s \in W_{3/2}^{(2)}(Y)$. Consequently $\nabla s \in \mathbf{W}_{3/2}^{(1)}(Y) \subseteq \mathbf{L}^3(Y)$, and, since from Lemma 1.1 $\mathbf{u} \in \mathbf{H}^1(Y)$, it follows that $(\mathbf{u}\nabla s) \in L^2(Y)$; recalling (1.15), we finally get $\frac{\partial^2 s}{\partial x_n^2} \in L^2(Y)$.

The estimation (1.41) reduces inside the domain to (1.37). On the boundary the estimation of $\frac{\partial^2 s}{\partial x_i \partial x_j}$ with $(i,j) \neq (n,n)$ can be made analogously; finally, the estimation of $\frac{\partial^2 s}{\partial x_n^2}$ follows directly from (1.15).

The second part of the theorem can be proved by the technique used in Lemma 1.1: the pressure $p$, the solution of the Neumann problem given by (1.29) and (1.30), has the property $p \in H^3(\Omega)$, because, in agreement with (1.41), $(s+\tau_h)\mathbf{g} \in \mathbf{H}^2(\Omega)$. □

## 2. The Small Rayleigh Number Case. Asymptotic Expansions

Here we prove that for sufficiently low Rayleigh numbers uniqueness holds, and that the unique solution is analytic and regular with respect to the Rayleigh number. Following on from these general results, we use the perturbation method to investigate a domain confined between two concentric spheres maintained at different uniform temperatures. Using the first three terms of the asymptotic expansions, two special values of the Rayleigh number are derived which characterize different types of flows—with one, two or three cells.

Recalling the Friedrichs inequality (1.25), we can specify what we meant previously by sufficiently low Rayleigh numbers:

$$a < a_0 \,. \tag{2.1}$$

The relation (2.1) provides the basis for this section, because of the following result:

THEOREM 2.1. *If* (2.1) *holds, then the solution of the problem* (1.19) *is unique.*

*Proof.* Let $(\mathbf{u}_1, s_1)$ and $(\mathbf{u}_2, s_2)$ be solutions of (1.19). Defining $\mathbf{u} = \mathbf{u}_1 - \mathbf{u}_2$ and $s = s_1 - s_2$, and subtracting the corresponding relations (1.20) and (1.21), we get

$$B(\mathbf{u},\mathbf{v}) = a(s, \mathbf{g}\cdot\mathbf{v}), \quad \forall \mathbf{v} \in \mathbf{H}\,, \tag{2.2}$$

$$A(s,t) + b(\mathbf{u}, s_2 + \tau_h, t) + b(\mathbf{u}_1, s, t) = 0, \quad \forall t \in H_0^1(\Omega)\,. \tag{2.3}$$

Taking $\mathbf{v} = \mathbf{u}$ in (2.2) and $t = s$ in (2.3), it is easy to obtain the estimations

$$|\mathbf{u}| \leqslant a|s|\,, \tag{2.4}$$

$$\|s\| \leqslant \frac{1}{2}|\mathbf{u}|\,. \tag{2.5}$$

From (2.4) and (2.5), with (1.25), it follows that

$$(1 - \frac{a}{a_0}) \| s \| \leqslant 0 ; \tag{2.6}$$

that is, because of (2.1), $\|s\| = 0$; finally (2.4) together with (1.25) implies that $|\mathbf{u}| = 0$. □

Another important point is that, if (2.1) holds, we can fix $\tau_h \in H^2(\Omega)$, which satisfies the properties of Lemma 5.1.1 in Chapter 1, because in this case $h_0 = 2a_0^{-2}(a_0 - 1)$ satisfies the condition (1.27) for any $a \in (0, a_0)$. Introducing $w = \tau_{h_0}$ and, with $(\mathbf{u}(a), s(a)) \in E_0$,

$$E_0 = (\mathbf{H} \times H_0^1(\Omega)) \cap (\mathbf{H}^2(\Omega) \times H^2(\Omega)),$$

which is the unique solution of (1.19); the system given by (1.20) and (1.21) then takes the form

$$B(\mathbf{u}(a), \mathbf{v}) = a(s(a) + w, \mathbf{g}\cdot\mathbf{v}), \quad \forall \mathbf{v} \in \mathbf{H}, \tag{2.7}$$

$$A(s(a), t) + b(\mathbf{u}(a), s(a) + w, t) = (\operatorname{div}(\mathbf{A}\nabla w), t), \quad \forall t \in H_0^1(\Omega). \tag{2.8}$$

PROPOSITION 2.1. $(\mathbf{u}(a+h), s(a+h)) \to (\mathbf{u}(a), s(a))$ *strongly in* $\mathbf{H} \times H_0^1(\Omega)$, *as* $h \to 0$.

*Proof.* From (2.7) and (2.8), with $\mathbf{v} = \mathbf{u}(a+h) - \mathbf{u}(a)$ and $t = s(a+h) - s(a)$, we have

$$|\mathbf{u}(a+h) - \mathbf{u}(a)| \leqslant \frac{2(a+h)}{a_0} \| s(a+h) - s(a) \| + \frac{1}{2} h,$$

$$\| s(a+h) - s(a) \| \leqslant \frac{1}{2} |\mathbf{u}(a+h) - \mathbf{u}(a)|,$$

and the proposition follows immediately. □

For any $a \in (0, a_0)$ and any $h \neq 0$ with the property $(a+h) \in (0, a_0)$, we define $D_h\mathbf{u}(a) = h^{-1}(\mathbf{u}(a+h) - \mathbf{u}(a))$ and $D_h s(a) = h^{-1}(s(a+h) - s(a))$.

In order to examine the analytic nature of $(\mathbf{u}(a), s(a))$, we differentiate formally (2.7) and (2.8) with respect to $a$, defining $(\mathbf{u}'(a), s'(a)) \in E_0$ by:

$$B(\mathbf{u}'(a), \mathbf{v}) = a(s'(a), \mathbf{g}\cdot\mathbf{v}) + (s(a) + w, \mathbf{g}\cdot\mathbf{v}), \quad \forall \mathbf{v} \in \mathbf{H}, \tag{2.9}$$

$$A(s'(a), t) + b(\mathbf{u}'(a), s(a) + w, t) + b(\mathbf{u}(a), s'(a), t) = 0, \forall t \in H_0^1(\Omega). \tag{2.10}$$

REMARK 2.1. The existence and uniqueness of $(\mathbf{u}'(a), s'(a))$ in $\mathbf{H} \times H_0^1(\Omega)$ can be proved as for Theorems 1.2 and 2.1, while the regularity property follows from arguments used in the previous section.

PROPOSITION 2.2. $(D_h\mathbf{u}(a), D_h s(a)) \to (\mathbf{u}'(a), s'(a))$ *strongly in* $\mathbf{H} \times H_0^1(\Omega)$ *as* $h \to 0$.

*Proof.* From (2.7) and (2.8), for any $(\mathbf{v}, t) \in \mathbf{H} \times H_0^1(\Omega)$ it follows that

$$B(D_h\mathbf{u}(a), \mathbf{v}) = a(D_h s(a), \mathbf{g}\cdot\mathbf{v}) + (s(a+h) + w, \mathbf{g}\cdot\mathbf{v}), \tag{2.11}$$

$$A(D_h s(a), t) + b(D_h \mathbf{u}(a), s(a+h) + w, t) +$$
$$+ b(\mathbf{u}(a), D_h s(a), t) = 0. \tag{2.12}$$

Subtracting (2.11) from (2.9) and (2.12) from (2.10) gives

$$B(D_h \mathbf{u}(a) - \mathbf{u}'(a), \mathbf{v}) = a(D_h s(a) - s'(a), \mathbf{g}\cdot\mathbf{v}) +$$
$$+ (s(a+h) - s(a), \mathbf{g}\cdot\mathbf{v}), \tag{2.13}$$

$$A(D_h s(a) - s'(a), t) + b(\mathbf{u}(a), D_h s(a) - s'(a), t) =$$
$$= b(D_h \mathbf{u}(a) - \mathbf{u}'(a), t, s(a+h) + w) -$$
$$- b(\mathbf{u}'(a), s(a+h) - s(a), t). \tag{2.14}$$

Taking $\mathbf{v} = D_h\mathbf{u}(a) - \mathbf{u}'(a)$ in (2.13) and $t = D_h s(a) - s'(a)$ in (2.14), we get the estimations

$$| D_h \mathbf{u}(a) - \mathbf{u}'(a)| \leqslant \frac{2a}{a_0} \| D_h s(a) - s'(a) \| + | s(a+h) - s(a)|, \tag{2.15}$$

$$\| D_h s(a) - s'(a) \| \leqslant \frac{1}{2} | D_h \mathbf{u}(a) - \mathbf{u}'(a)| +$$

$$+ | u'(a)|_4 | s(a+h) - s(a)|_4 . \tag{2.16}$$

If we use (2.15) and (2.16) together with Proposition 2.1, the proof is completed. □

Introducing $(\mathbf{u}^{(0)}(a), s^{(0)}(a)) = (\mathbf{u}(a), s(a))$, $(\mathbf{u}^{(1)}(a), s^{(1)}(a)) = (\mathbf{u}'(a), s'(a))$, we define recursively for any $m \geqslant 2$, $(\mathbf{u}^{(m)}(a), s^{(m)}(a)) \in E_0$:

$$B(\mathbf{u}^{(m)}(a), \mathbf{v}) = (as^{(m)}(a) + ms^{(m-1)}(a), \mathbf{g}\cdot\mathbf{v}), \quad \forall \mathbf{v} \in \mathbf{H}, \tag{2.17}$$

$$A(s^{(m)}(a), t) + b(\mathbf{u}^{(m)}(a), w, t) +$$

$$+ \sum_{k=0}^{m} \binom{m}{k} b(\mathbf{u}^{(m-k)}(a), s^{(k)}(a), t) = 0, \quad \forall t \in H_0^1(\Omega). \tag{2.18}$$

As in the proofs of Propositions 2.1 and 2.2, one can prove by complete induction the propositions

$P_1(m): (\mathbf{u}^{(m-1)}(a+h), s^{(m-1)}(a+h)) \to_{(h\to 0)} (\mathbf{u}^{(m-1)}(a), s^{(m-1)}(a))$

strongly in $\mathbf{H} \times H_0^1(\Omega)$.

$P_2(m): (D_h \mathbf{u}^{(m-1)}(a), D_h s^{(m-1)}(a)) \to_{(h\to 0)} (\mathbf{u}^{(m)}(a), s^{(m)}(a))$

strongly in $\mathbf{H} \times H_0^1(\Omega)$.

Thus we have proved:

PROPOSITION 2.3. *The solution of the problem* (1.19) *is of class* $C^\infty$ *with respect to any* $a \in (0, a_0)$. □

We reconsider now (2.17) and (2.18) by setting $\mathbf{v} = \mathbf{u}^{(m)}(a)$ and $t = s^{(m)}(a)$; it follows that

$$|\mathbf{u}^{(m)}(a)| \leqslant \frac{2a}{a_0}\| s^{(m)}(a) \| + \frac{2m}{a_0} \| s^{(m-1)}(a) \| , \tag{2.19}$$

$$\| s^{(m)}(a) \| \leqslant \frac{1}{2} |\mathbf{u}^{(m)}(a)| + \sum_{k=1}^{m-1} \binom{m}{k} |\mathbf{u}^{(m-k)}(a)|_4 | s^{(k)}(a)|_4 . \tag{2.20}$$

Passing on to the Neumann problem of the pressure, equivalent to (2.17), as in Lemma 1.3 we get

$$\|\mathbf{u}^{(m)}(a) \|_1 \leqslant c_0 \| as^{(m)}(a) + ms^{(m-1)}(a) \| , \quad \forall m \geqslant 1 . \tag{2.21}$$

Thus, from (2.19) and (2.20), we have

$$\| s^{(m)}(a) \| \leqslant \frac{a}{a_0} \| s^{(m)}(a) \| + \frac{m}{a_0} \| s^{(m-1)}(a) \| +$$

$$+ c_1 \sum_{k=1}^{m-1} \binom{m}{k} \| s^{(k)}(a) \| ( \| s^{(m-k)}(a) \| + (m-k) \| s^{(m-k-1)}(a) \| ) . \tag{2.22}$$

Equation (2.22) is easily reduced to the estimation

$$\frac{1}{m!} \| s^{(m)}(a) \| \leqslant A_m , \quad \forall m \geqslant 1 , \tag{2.23}$$

where $A_m$ is the sequence defined by

$$A_m = c_2 \sum_{k=1}^{m-1} A_k A_{m-k} , \quad \forall m \geqslant 2, A_1 = \|s'(a) \| \tag{2.24}$$

(everywhere $c_i > 0$ ($i = 0,1,2, \ldots$) denote certain constants). As $A_m$ can be determined from (2.24),

$$A_m = \frac{1}{m} \binom{2m-2}{m-1} c_2^{m-1} \| s'(a) \|^m , \quad \forall m \geqslant 1 , \tag{2.25}$$

and from (2.9) and (2.10), using the standard estimation method, we obtain

$$\| s'(a) \| \leqslant \frac{a_0}{4 (a_0 - a)} . \tag{2.26}$$

It follows that

$$\frac{1}{m!} \| s^{(m)} (a) \| \leqslant (c_3 (a_0 - a))^{-m} . \tag{2.27}$$

With the comparison test we have finally

PROPOSITION 2.4. *The Taylor's series*

$$\mathscr{T}(a, h) = \sum_{m \geq 0} \frac{h^m}{m!} (\mathbf{u}^{(m)}(a), s^{(m)}(a)) := (\mathbf{U}(a, h), S(a, h)) \tag{2.28}$$

*is well defined in* $\mathbf{H} \cap \mathbf{H}^1(\Omega) \times H_0^1(\Omega)$; *moreover, its radius of convergence is at least equal to* $c_3(a_0 - a)$.

Our next objective is to show that $(\mathbf{u}(a), s(a))$ is regular. This is embodied in the following proposition:

PROPOSITION 2.5. $\mathscr{T}(a, h) = (\mathbf{u}(a+h), s(a+h))$ *for any* $h \in \mathbb{R}$ *with* $|h| < c_3(a_0 - a)$.

*Proof.* Defining the partial sums

$$\mathbf{U}_m(a, h) = \sum_{n=0}^{m} \frac{h^n}{n!} \mathbf{u}^{(n)}(a), \qquad S_m(a, h) = \sum_{n=0}^{m} s^{(n)}(a) \frac{h^n}{n!}, \tag{2.29}$$

and using (2.7), (2.9) and (2.17), we get

$$B(\mathbf{U}_m(a, h), \mathbf{v}) = (a+h)(S_m(a+h) + w, \mathbf{g}\cdot\mathbf{v}) - \\ - \frac{h^{m+1}}{m!}(s^{(m)}(a), \mathbf{g}\cdot\mathbf{v}), \quad \forall \mathbf{v} \in \mathbf{H}\,. \tag{2.30}$$

Passing (2.30) to the limit, using Proposition 2.4 and the relation (2.27) it follows that

$$B(\mathbf{U}(a, h), \mathbf{v}) = (a+h)(S(a+h) + \mathbf{w}, \mathbf{g}\cdot\mathbf{v}), \quad \forall\, \mathbf{v} \in \mathbf{H}\,. \tag{2.31}$$

We treat the second equation of the system in the same way: with (2.8), (2.10) and (2.18), we get

$$A(S_m(a, h), t) + b(\mathbf{U}_m(a, h), S_m(a, h) + w, t) = (\operatorname{div}(\mathbf{A}\nabla w), t) + \\ + \sum_{n=1}^{m} \sum_{k=0}^{m-n} b\left(\frac{h^{n+k}}{(n+k)!}\mathbf{u}^{(n+k)}(a), \frac{h^{m-k}}{(m-k)!} s^{(m-k)}(a), t\right), \forall t \in H_0^1(\Omega) \tag{2.32}$$

and, as the last term of (2.32) is bounded by

$$\left(\sum_{n=1}^{m} \sum_{k=0}^{m-n} \rho^{m+n}\right) \|t\| \quad \text{with some } \rho \in (0,1),$$

passing (2.32) to the limit we obtain

$$A(S(a, h), t) + b(\mathbf{U}(a, h), S(a, h) + w, t) = (\operatorname{div}(\mathbf{A}\nabla w), t), \\ \forall t \in H_0^1(\Omega)\,. \tag{2.33}$$

Since the system given by (2.31) and (2.33) is identical to the system given by (2.7) and (2.8), which has a unique solution (Theorem 2.1), the proof is completed.

It is now obvious that with Propositions 2.3, 2.4 and 2.5, we have proved the following theorem:

THEOREM 2.2. *The solution of the problem* (1.19) *is analytic and regular with respect to any* $a \in (0, a_0)$. □

The following results are to be considered as complementary to the work by Burns and Tien (1979) and Poliševski (1979), in which the problem of natural convection in a fluid-saturated porous medium contained between two impervious concentric spheres was studied by straightforward expansions in terms of the Rayleigh number. We start by recalling this problem.

The spheres have radii $R_1$, $R_2$ and ($R_1 < R_2$) and are centred on 0, the origin of the coordinate system $Ox_1\ x_2\ x_3$, negative $Ox_3$ direction is the direction in which the gravitational force acts ($\mathbf{g} = -g\,\mathbf{e}_3$, $g > 0$). We adopt here a particular form of the system (1.1)–(1.3) (in fact the same classical form as used by Ene and Gogonea (1973)):

$$\operatorname{div} \mathbf{u} = 0\,, \tag{2.34}$$

$$\frac{\mu}{k}\mathbf{u} + \nabla p = \rho_f\,(1 - \alpha\,(T - T_0))\,\mathbf{g}\,, \tag{2.35}$$

$$\rho_f\, c_f \mathbf{u}\, \nabla T = \chi\, \Delta T\,, \tag{2.36}$$

for any $r = \sqrt{x_1^2 + x_2^2 + x_3^2} \in (R_1, R_2)$, where $\alpha$, $c_f$, $\mu$ and $\rho_f$ are the coefficient of thermal expansion, the specific heat (at constant volume), the viscosity and the density of the saturating fluid, and $k$ and $\chi$ are the permeability and the thermal conductivity of the porous medium.

Assuming that heating or cooling occurs uniformly along the boundaries, the system (2.34)–(2.36) must be solved subject the following conditions:

$$\mathbf{u} \cdot \mathbf{v} = 0, \quad r = R_1 \text{ and } r = R_2\,, \tag{2.37}$$

$$T = T_1\,, \quad r = R_1\,, \tag{2.38}$$

$$T = T_2\,, \quad r = R_2\,. \tag{2.39}$$

We assume that $T_1 \neq T_2$ because if $T_1 = T_2$ the problem is trivial. We now consider a dimensionless from of the problem:

$$x^* = R_1^{-1}x \qquad q^* = R_1^{-1}R_2 > 1\,,$$
$$T^* = (T_1 - T_2)^{-1}(T - T_2) \qquad u^* = c_f\, R_1\, \chi^{-1} u\,,$$
$$h^* = \mu^{-1}\chi^{-1}\rho_f\, c_f\, k\,(p + g(1 + \frac{1}{2}\alpha\,(T_0 - T_2))\,x_3)$$
$$a^* = \mu^{-1}\chi^{-1}\rho_f^2\, c_f\, R_1 k g \alpha\,(T_1 - T_2)\,.$$

REMARK 2.2. The Rayleigh number $a^*$ covers the case $T_1 > T_2$ (with $a^* > 0$) as well as the case $T_1 < T_2$ (with $a^* < 0$).

Omitting the asterisks to simplify the notation we obtain

$$\operatorname{div} \mathbf{u} = 0, \qquad r \in (1, q), \tag{2.40}$$

$$\mathbf{u} + \nabla h = aT\mathbf{e}_3, \qquad r \in (1, q), \tag{2.41}$$

$$\mathbf{u}\nabla T = \Delta T, \qquad r \in (1, q), \tag{2.42}$$

$$\mathbf{u}\cdot\mathbf{v} = 0, \qquad r = 1, r = q, \tag{2.43}$$

$$T = 1, \qquad r = 1, \tag{2.44}$$

$$T = 0, \qquad r = q. \tag{2.45}$$

Changing to spherical polar coordinates $(r, \theta, \varphi)$ the problem (2.40)–(2.45) becomes

$$\frac{\partial}{\partial r}(r^2 \sin\theta\, u_r) + \frac{\partial}{\partial\theta}(r\sin\theta\, u_\theta) + \frac{\partial}{\partial\varphi}(r u_\varphi) = 0, r \in (1, q), \tag{2.46}$$

$$u_r + \frac{\partial h}{\partial r} = aT\cos\theta, \qquad r \in (1, q), \tag{2.47}$$

$$u_\theta + \frac{1}{r}\frac{\partial h}{\partial\theta} = -aT\sin\theta, \qquad r \in (1, q), \tag{2.48}$$

$$u_\varphi + \frac{1}{r\sin\theta}\frac{\partial h}{\partial\varphi} = 0, \qquad r \in (1, q), \tag{2.49}$$

$$\frac{1}{r^2\sin\theta}\left(\frac{\partial}{\partial r}\left(r^2\sin\theta\frac{\partial T}{\partial r}\right) + \frac{\partial}{\partial\theta}\left(\sin\theta\frac{\partial T}{\partial\theta}\right) + \frac{\partial}{\partial\varphi}\left(\frac{1}{\sin\theta}\frac{\partial T}{\partial\varphi}\right)\right) =$$

$$= u_r\frac{\partial T}{\partial r} + \frac{u_\theta}{r}\frac{\partial T}{\partial\theta} + \frac{u_\varphi}{r\sin\theta}\frac{\partial T}{\partial\varphi}, \quad r \in (1, q), \tag{2.50}$$

$$u_r = 0, \qquad r = 1, \tag{2.51}$$

$$u_r = 0, \qquad r = q, \tag{2.52}$$

$$T = 1, \qquad r = 1, \tag{2.53}$$

$$T = 0, \qquad r = q. \tag{2.54}$$

PROPOSITION 2.6. *For any $a \neq 0$, $q > 1$, the total heat transfer is different from the heat transfer which occurs by pure conduction alone.*

*Proof.* Suppose that there exists $a \neq 0$ and $q > 1$ for which $\mathbf{u}\nabla T = 0$; then from Equation (2.50) together with the boundary conditions (2.53) and (2.54) we find that

$$T = \frac{1}{q-1}\left(\frac{q}{r} - 1\right). \tag{2.55}$$

The condition $\mathbf{u}\nabla T = 0$ can be shown to reduce to $u_r = 0$, and thus from (2.46)–(2.54) it remains to verify

$$\frac{\partial}{\partial\theta}(\sin\theta\, u_\theta) + \frac{\partial u_\varphi}{\partial\varphi} = 0\,, \tag{2.56}$$

$$\frac{\partial h}{\partial r} = \frac{a}{q-1}\left(\frac{q}{r} - 1\right)\cos\theta\,, \tag{2.57}$$

$$u_\theta + \frac{1}{r}\frac{\partial h}{\partial\theta} = \frac{-a}{q-1}\left(\frac{q}{r} - 1\right)\sin\theta\,, \tag{2.58}$$

$$u_\varphi + \frac{1}{r\sin\theta}\frac{\partial h}{\partial\varphi} = 0 \tag{2.59}$$

for any $r\in(1, q)$. From (2.57), we have

$$h = \frac{a}{q-1}((\mathrm{q}\ln(r) - r)\cos\theta + qf(\theta,\varphi)), \tag{2.60}$$

where $f=f(\theta, \varphi)$, via the elimination of $u_\theta$ and $u_\varphi$ between (2.56), (2.58) and (2.59), have to verify

$$\frac{\partial}{\partial\theta}(\sin\theta\frac{\partial f}{\partial\theta}) + \frac{\partial^2 f}{\partial\varphi^2} = (-1 + \ln r)\sin 2\theta \tag{2.61}$$

which is obviously impossible. □

We now look for asymptotic solution of the form

$$(\mathbf{u}, h, T) = \sum_{m\geqslant 0}(\mathbf{u}_m, h_m, T_m)\frac{a^m}{m!}\,, \tag{2.62}$$

where the $a$-independent functions $(\mathbf{u}_m, h_m, T_m)$ are assumed to satisfy the set of equations obtained by equating the coefficients of the different powers of $a$ which occur in the formal expansion of the system (2.46)–(2.54). It is then easily verified that (2.62) are the Taylor series around $a=0$, so we can make use of previous results obtained in this section. We know that the successive problems (2.17)–(2.18) have unique solutions; that is why in the present case they are independent of $\varphi$. This fact, together with (2.46), allows us to introduce a stream function $\Psi$ given by

$$r^2\sin\theta\, u_r = \frac{\partial\Psi}{\partial\theta}\,, \qquad r\sin\theta\, u_\theta = -\frac{\partial\Psi}{\partial r}\,. \tag{2.63}$$

Thus, eliminating $h$, the system (2.46)–(2.54) becomes

$$\frac{1}{\sin\theta}\frac{\partial^2\Psi}{\partial r^2} + \frac{1}{r^2}\frac{\partial}{\partial\theta}\left(\frac{1}{\sin\theta}\frac{\partial\Psi}{\partial\theta}\right) = a\left(\cos\theta\frac{\partial T}{\partial\theta} + r\sin\theta\frac{\partial T}{\partial r}\right), r\in(1,q), \tag{2.64}$$

$$\sin\theta\frac{\partial}{\partial r}\left(r^2\frac{\partial T}{\partial r}\right) + \frac{\partial}{\partial\theta}\left(\sin\theta\frac{\partial T}{\partial\theta}\right) = \frac{\partial T}{\partial r}\frac{\partial\Psi}{\partial\theta} - \frac{\partial T}{\partial\theta}\frac{\partial\Psi}{\partial r}\,,\quad r\in(1, q)\,, \tag{2.65}$$

$$T = 1 \text{ and } \Psi = 0,\quad r=1\,, \tag{2.66}$$

$$T = 0 \text{ and } \Psi = 0,\quad r=q\,. \tag{2.67}$$

We emphasize that in introducing the stream function $\Psi$ we are assuming that the unknowns have the form (2.62). In general, we do not know whether ($\mathbf{u}$, $h$, $T$) are $\varphi$-independent.

Taking $(\Psi, T)$ in the form

$$(\Psi, T) = \sum_{m \geq 0} (\Psi_m, T_m)\, a^m, \tag{2.68}$$

$(\Psi_m, T_m)$ can be defined recursively, for any $m \geq 1$, by

$$r^2 \frac{\partial^2 \Psi_m}{\partial r^2} + \sin\theta \frac{\partial}{\partial\theta}\left(\frac{1}{\sin\theta}\frac{\partial \Psi_m}{\partial\theta}\right) = r^2 \sin\theta \left(\cos\theta \frac{\partial T_{m-1}}{\partial\theta} - \right.$$

$$\left. - r\sin\theta \frac{\partial T_{m-1}}{\partial r}\right), \quad r \in (1, q), \tag{2.69}$$

$$\Psi_m = 0, \quad r = 1, r = q, \tag{2.70}$$

$$\sin\theta \frac{\partial}{\partial r}\left(r^2 \frac{\partial T_m}{\partial r}\right) + \frac{\partial}{\partial\theta}\left(\sin\theta \frac{\partial T_m}{\partial\theta}\right) = \sum_{k=1}^{m}\left(\frac{\partial T_{m-k}}{\partial r}\frac{\partial \Psi_k}{\partial\theta} - \frac{\partial T_{m-k}}{\partial\theta}\frac{\partial \Psi_k}{\partial r}\right),$$

$$r \in (1, q), \tag{2.71}$$

$$T_m = 0, \quad r = 1, r = q, \tag{2.72}$$

where the first terms are

$$\Psi_0 = 0, \quad T_0 = \frac{1}{q-1}\left(\frac{q}{r} - 1\right). \tag{2.73}$$

The expressions for $(\Psi_1, T_1)$ and $(\Psi_2, T_2)$, as we have mentioned, can be found in Burns and Tien (1979) and Poliševski (1979) together with various discussions (on the configurations of the streamlines and isothermal lines, the Nusselt number, and so on). Here we must emphasize two points. First, if we let $q \to \infty$, then we find the results obtained by Yamamoto (1974) as a limiting case of the present problem; that is, the problem of natural convection about a heated sphere with the boundary condition $\Psi = 0$, $T = 1$ as $r = 1$ and $|\mathbf{u}| \to 0$, $T \to 0$ as $r \to \infty$. Second, a multicellular flow can be obtained even with its second order approximation

$$\Psi \simeq a\Psi_1 + a^2\Psi_2. \tag{2.74}$$

If we use (2.74) to find the streamlines on which $\Psi = 0$, we get, as well as

$$r = q, \qquad r = 1 \qquad \theta = 0 \qquad \theta = \pi, \tag{2.75}$$

the $(r, \theta)$ points which satisfy the equation

$$\left(\frac{q+1}{q} + \frac{1}{r}\right) + \frac{a\cos\theta}{12\,(q^5-1)} H(r, q) = 0, \tag{2.76}$$

where

$$H(r, q) = r^{-2} q^2 (q+1)(q^2+q+1) - r^{-1} q(5q^4 + 3q^3 + 2q^2 + 3q + 5) +$$

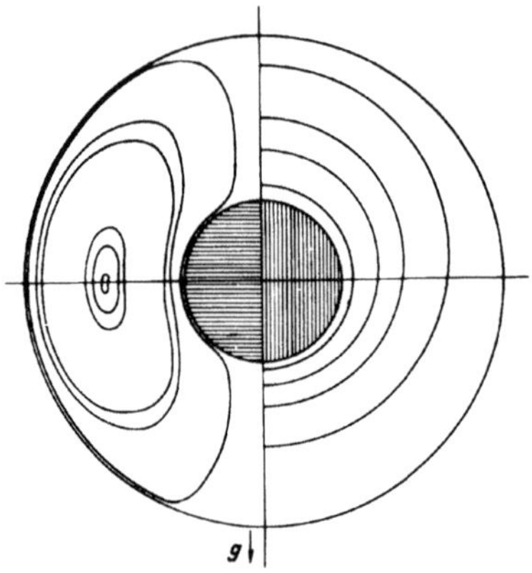


Fig. 6. Temperature distribution and flow pattern for $a \in (0, a_1)$.

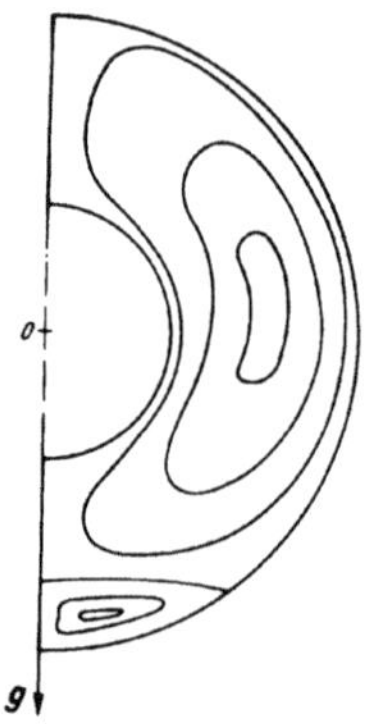


Fig. 7. Streamline pattern for $a \in (a_1, a_2)$.

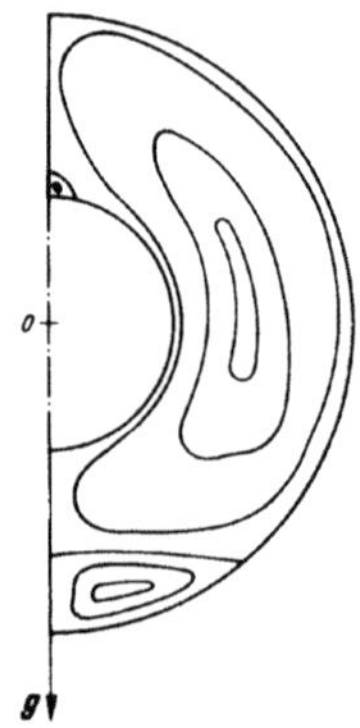


Fig. 8. Streamline pattern for $a \in (a_2, \infty)$.

$$+ (q^2+q+1)^{-1}((q+1)(4q^6+6q^4+7q^3+6q^2+4) + r(q^6-6q^5-$$
$$- 3q^4-2q^3-3q^2-6q+1)).$$

We introduce the parameters

$$a_1 = \frac{12(q^5-1)(q+2)}{qH(q,q)} > 0, \quad a_2 = \frac{q^2(2q+1)}{(q+2)} a_1 > a_1$$

(for instance when $q = 2.5$, $a_1 \simeq 17$ and $a_2 \simeq 142$); then, for $T_1 > T_2$,

(a) for $a \in (0, a_1)$, (2.76) has no solution; this is the usual unicellular flow (Fig. 6);

(b) for $a \in (a_1, a_2)$ there appears in the relatively stagnant and stable cold region at the bottom of the enclosure a second cell (Fig. 7);

(c) for $a \in (a_2, \infty)$ a third small cell appears at the top of the inner sphere (Fig. 8).

If $T_1 < T_2$ the discussion is somewhat similar, the configurations being reversed in comparison with the previous cases.

Experimentalists (see for instance J. Fauveau *et al.* (1978)) have mentioned the small cell which appears at the top of the inner sphere, but no distinct motion has yet been observed at the bottom of the enclosure.

## 3. The Structure of the Steady Solutions

Here we deal with some general and generic properties of a set of solutions analogous to that obtained for the Navier – Stokes equations (Foiaş and Temam (1977)). First we prove that the set of solutions is homeomorphic with a compact set of $\mathbb{R}^m$ ($m$ sufficiently large), emphasizing the importance of temperature in this problem. We then show that for almost all data the set of solutions is finite, while the continuum of the solutions (with respect to the Rayleigh number) is an one-dimensional manifold of class $C^1$.

We introduce $A_0 : \mathscr{D}(\Omega) \to L^2(\Omega)$, given by

$$A_0(T) = -\operatorname{div}(\mathbf{A}\nabla T) = -A_{ij}\frac{\partial^2 T}{\partial x_i \partial x_j},$$

this operator is positive-defined and self-adjoint in the sense of Lagrange. Consequently, it admits a self-adjoint extension $\tilde{A}_0$ with the properties

(a) the domain of $\tilde{A}_0$ is given by $D(\tilde{A}_0) = D(A_0^*) \cap H_0^1(\Omega)$;

(b) the range of $\tilde{A}_0$ is given by $R(\tilde{A}_0) = L^2(\Omega)$;

(c) $(\tilde{A}_0(S), S) \geqslant \|S\|^2, \quad \forall S \in D(\tilde{A}_0)$.

As $\tilde{A}_0^{-1} : L^2(\Omega) \to L^2(\Omega)$ is a self-adjoint operator, from property (a), via Rellich's lemma, it follows that $\tilde{A}_0^{-1}$ is compact too. Hence, there exists a sequence $\{r_k\}_k$ of orthonormal eigenvectors of $\tilde{A}_0^{-1}$ which form an orthonormal basis in $L^2(\Omega)$; also, the eigenvalues $\lambda_k^{-1} \to 0$ of $\tilde{A}_0^{-1}$ have the property

$$1 \leqslant \lambda_1 \leqslant \lambda_2 \leqslant \cdots \leqslant \lambda_k \leqslant \cdots \to \infty .$$

Since $\tilde{A}_0^{-1} r_k = \lambda_k^{-1} r_k$ is equivalent to $\tilde{A}_0 r_k = \lambda_k r_k$, $\{\lambda_k\}_k$ are the eigenvalues of $\tilde{A}_0$, corresponding to the same eigenvectors.

LEMMA 3.1. $A(r_k, T) = \lambda_k(r_k, T), \quad \forall T \in H_0^1(\Omega)$.

*Proof.* For any $T \in \mathscr{D}(\Omega)$, we have

$$A(r_k, T) = \int_\Omega A_{ij}\frac{\partial T}{\partial x_i \partial x_j}\frac{\partial r_k}{\partial x_j} = -\int_\Omega r_k \frac{\partial}{\partial x_j}\left(A_{ij}\frac{\partial T}{\partial x_i}\right) =$$

$$= -\int_\Omega r_k \frac{\partial}{\partial x_i}\left(A_{ij}\frac{\partial T}{\partial x_j}\right) = (r_k, A_0(T)) = (A_0^*(r_k), T). \tag{3.1}$$

Because $\tilde{A}_0 = A_0^* \mid D(A_0^*) \cap H_0^1(\Omega)$, (3.1) becomes

$$A(r_k, T) = \lambda_k(r_k, T), \quad \forall T \in \mathscr{D}(\Omega)$$

and, as $\mathscr{D}(\Omega)$ is dense in $H_0^1(\Omega)$, the relation follows by continuity. □

LEMMA 3.2. $|Q_m S| \leqslant C\lambda_{m+1}^{-1/2}\|Q_m S\|$, $\forall S \in H_0^1(\Omega)$, *where* $Q_m = I - P_m$, $P_m$ *being the orthogonal projection onto the space spanned by* $r_1, \cdots, r_m$.

*Proof.* Let

$$Q_m S = \sum_{i \geqslant m+1} \sigma_i r_i .$$

It then follows that

$$A(Q_m S, Q_m S) = \sum_{i \geqslant m+1} \sigma_i A(r_i, Q_m S) = \sum_{i \geqslant m+1} \sigma_i \lambda_i (r_i, Q_m S) =$$

$$= \sum_{i \geqslant m+1} \lambda_i \sigma_i^2 \geqslant \lambda_{m+1} \sum_{i \geqslant m+1} \sigma_i^2 = \lambda_{m+1} | Q_m S|^2 ;$$

on the other hand, evidently $A(Q_m S, Q_m S) \leqslant C \| Q_m S \|^2$. □

LEMMA 3.3. $\lambda_m \geqslant C(\Omega)\, m^{2/3}$.

*Proof.* This is easily proved using an estimation given by Agmon (1965, Sect. 13):

$$| S(x) | \leqslant C_1 \| S \|_2^{3/4} | S |^{1/4} \quad \text{a.e. on } \Omega, \forall S \in H^2(\Omega). \tag{3.2}$$

Let $\sigma_i \in \mathbb{R}$, $i \in \{1, 2, \cdots, m\}$, and let $S(x) = \sum_{i=1}^{m} \sigma_i r_i(x)$ in (3.2); it then follows that

$$\left| \sum_{i=1}^{m} \sigma_i r_i(x) \right| \leqslant C_1 \left\| \sum_{i=1}^{m} \sigma_i r_i \right\|_2^{3/4} \left| \sum_{i=1}^{m} \sigma_i r_i \right|^{1/4}. \tag{3.3}$$

From the classical regularity result for the Dirichlet problem (see Lions and Magenes (1968), Chap. 2) we have

$$\| S \|_2 \leqslant C_2 \left| \tilde{A}_0(S) \right|, \quad \forall S \in D(\tilde{A}_0), \tag{3.4}$$

and with the same $S$ as in (3.3) we get

$$\left\| \sum_{i=1}^{m} \sigma_i r_i \right\|_2 \leqslant C_2 \left| \tilde{A}_0 \left( \sum_{i=1}^{m} \sigma_i r_i \right) \right| = C_2 \left| \sum_{i=1}^{m} \sigma_i \lambda_i r_i \right| \leqslant$$

$$\leqslant C_2 \lambda_m \left| \sum_{i=1}^{m} \sigma_i r_i \right| . \tag{3.5}$$

Introducing (3.5) in (3.3), we have

$$\left| \sum_{i=1}^{m} \sigma_i r_i(x) \right| \leqslant C_3 \lambda_m^{3/4} \left| \sum_{i=1}^{m} \sigma_i r_i \right| = C_3 \lambda_m^{3/4} \left( \sum_{i=1}^{m} \sigma_i^2 \right)^{1/2} .$$

Thus we obtain the relation

$$\sum_{i=1}^{m} | r_i(x) |^2 \leqslant C_4 \lambda_m^{3/2}$$

which, by integration, and using the fact that $|r_i| = 1\ \forall i$, completes the proof. □

We denote by $\mathcal{S}$ the set of solutions of the problem (1.19), and by $F_m: \mathcal{S} \to L^2(\Omega)$ the operator $F_m(\mathbf{u}, s) = P_m s$, which is obviously continuous if $\mathcal{S}$ is endowed with the topology of $\mathbf{H} \times H_0^1(\Omega)$.

LEMMA 3.4. *$F_m$ is injective if $m$ is sufficiently large.*

*Proof.* Let $(\mathbf{u}_1, s_1), (\mathbf{u}_2, s_2)$ be from $\mathcal{S}$ and let $(\mathbf{u}, s) = (\mathbf{u}_1 - \mathbf{u}_2, s_1 - s_2)$. From (1.33) we have

$$B(\mathbf{u}, \mathbf{v}) + (as, \mathbf{g} \cdot \mathbf{v}) = 0, \quad \forall\, \mathbf{v} \in \mathbf{H}, \tag{3.6}$$

$$A(s, t) + b(\mathbf{u}, s_2 + \tau_h, t) + b(\mathbf{u}_1, s, t) = 0, \quad \forall t \in H_0^1(\Omega). \tag{3.7}$$

Choosing $t = Q_m s$ in (3.7) and using Theorem 1.4, we obtain

$$\|Q_m s\| \leqslant \frac{1}{2}|\mathbf{u}| + |\mathbf{u}_1|_\infty |P_m s| \leqslant \frac{1}{2}|\mathbf{u}| + C_4 |P_m s| \cdot \tag{3.8}$$

From (3.6) with $\mathbf{v} = \mathbf{u}$, we have

$$|\mathbf{u}| \leqslant a|s| \leqslant a|P_m s| + a|Q_m s|, \tag{3.9}$$

and the relation (3.8) becomes

$$\|Q_m s\| \leqslant \frac{1}{2} a |Q_m s| + (\frac{1}{2}a + C_4)|P_m s|. \tag{3.10}$$

Using Lemma 3.2, from (3.10) it follows that

$$(1 - \frac{1}{2} ac\, \lambda_{m+1}^{-1/2}) \|Q_m s\| \leqslant (\frac{1}{2}a + C_4)|P_m s|. \tag{3.11}$$

From Lemma 3.3, we find that for sufficiently large $m$ the relation

$$(1 - \frac{1}{2} ac\, \lambda_{m+1}^{-1/2}) > 0 \tag{3.12}$$

holds. Hence, if $|P_m s| = 0$, from (3.11) we have that $\|Q_m s\| = 0$; using Lemma 3.2 once more, it follows that $|Q_m s| = 0$, and finally, from (3.9), $|\mathbf{u}| = 0$. □

THEOREM 3.1. *$\mathscr{S} \subseteq \mathbf{H} \times H_0^1(\Omega)$ is homeomorphic to a compact set of $\mathbb{R}^m$, $m$ being sufficiently large for* (3.12) *to be satisfied.*

*Proof.* From Theorem 1.4 and from Sobolev's embedding theorems, $\mathscr{S}$ is compact in $\mathbf{H} \times H_0^1(\Omega)$. If $m$ satisfies (3.12), then $F_m(\mathscr{S})$ is compact in $\mathbb{R}^m$ and $F_m^{-1}$ is continuous on this set; this follows from (3.9), (3.11) and

$$\|P_m s\|^2 \leqslant A(P_m s, P_m s) \leqslant \lambda_m |P_m s|^2.$$ □

THEOREM 3.2. *$\mathscr{S}$ is compact in $E_0$ (defined after Theorem* 2.1).

*Proof.* Let $(\mathbf{u}_i, s_i)$ be a sequence of elements of $\mathscr{S} \subseteq E_0$. As $\{(\mathbf{u}_i, s_i)\}_i$ is bounded in $E_0$ (see Theorem 1.4), it contains a sub-sequence (still denoted by $i$) which is weakly convergent in $E_0$ and strongly convergent in $\mathbf{H} \cap \mathbf{H}^1(\Omega) \times H_0^1(\Omega)$ to some element $(\mathbf{u}_0, s_0) \in E_0$. Setting $(\mathbf{v}_i, t_i) = (\mathbf{u}_i - \mathbf{u}_0,\ s_i - s_0)$, the relation (3.7) becomes

$$\tilde{A}_0(t_i) + \mathbf{v}_i \nabla (s_i + \tau_h) + \mathbf{u}_0 \nabla t_i = 0 \quad \text{in } L^2(\Omega). \tag{3.13}$$

From (3.13) we obtain

$$|\tilde{A}_0(t_i)| \leqslant c_1 \|\mathbf{v}_i\|_1 \|s_i + \tau_h\|_2 + |\mathbf{u}_0|_\infty \|t_i\|. \tag{3.14}$$

Now, from (3.14), via (3.4), we can see that $\|t_i\|_2 \to 0$. Taking into account the corresponding pressures $p_i$ and $p_0$, then $q_i = p_i - p_0$ must satisfy the conditions

$$-\operatorname{div}(\mathbf{K} \nabla q_i) = a \operatorname{div}(\mathbf{K} \mathbf{g} t_i) \quad \text{in } \Omega, \tag{3.15}$$

$$(\mathbf{K} \nabla q_i)\, \mathbf{v} = 0 \quad \text{on } \partial\Omega. \tag{3.16}$$

From the regularity properties of the Neumann problem (see Lions and Magenes (1968), Chap. 2), we get

$$\|q_i\|_3 \leqslant c_2 \|a \operatorname{div}(\mathbf{K}\mathbf{g} t_i)\|_1 \leqslant c_3 \|t_i\|_2 \to 0 \cdot \tag{3.17}$$

The fact that $\|\mathbf{v}_i\|_2 \to 0$ now follows directly from (1.14). □

REMARK 3.1. (An implicit representation of the solutions). If $m$ satisfies (3.12), then Lemma 3.4 allows us to define for any $\xi \in P_m(H_0^1(\Omega) \cap H^2(\Omega))$ the unique element $(\mathbf{u}(\xi), \sigma(\xi)) \in \mathbf{H} \times Q_m(H_0^1(\Omega) \cap H^2(\Omega))$ which satisfies the conditions

$$P_H(\mathbf{B}\mathbf{u}(\xi) + a(\xi + \sigma(\xi) + \tau_h)\,\mathbf{g}) = 0\,, \tag{3.18}$$

$$-\operatorname{div}(\mathbf{A}\nabla\sigma(\xi)) + Q_m(\mathbf{u}(\xi)\nabla(\xi + \sigma(\xi) + \tau_h)) = $$
$$= \operatorname{div}(\mathbf{A}\nabla(\xi + Q_m\tau_h))\,, \tag{3.19}$$

where $P_H$ is the projection of $\mathbf{L}^2(\Omega)$ on $\mathbf{H}$.

Obviously, $(\mathbf{u}(\xi), \xi + \sigma(\xi))$ is a solution of our problem if and only if $\xi$ satisfies

$$-\operatorname{div}(\mathbf{A}\nabla\xi) + P_m(\mathbf{u}(\xi)\nabla(\xi + \sigma(\xi) + \tau_h)) = \operatorname{div}(\mathbf{A}\nabla P_m\tau_h)\,\cdot \tag{3.20}$$ □

The generic results given in the remainder of this section are based on Smale's density theorem in the improved form used by Abraham and Robbin (1967, Chap. 3). We start by recalling this theorem.

Let $E$ and $F$ be Banach spaces, and $L: E \to F$ a linear continuous operator; then $L$ is a Fredholm operator if

(a) the kernel of $L$ is finite-dimensional, and
(b) the image of $L$ is closed and has finite codimension.

The index of a Fredholm operator is defined by

$$\operatorname{ind} L = \dim(\ker L) - \dim(F/R(L)).$$

If $f: E \to F$ is a $C^1$ map, then $f$ is a Fredholm map if for every $x \in E$, $Df(x)$ is a Fredholm operator. For such an operator an index can be defined by ind $f=$ ind $Df(x)$, because ind $Df(x)$ is independent of $x$.

A point $x \in E$ is a regular point of $f$ if $Df(x)$ is surjective; otherwise, $x$ is a critical point of $f$.

The image of the critical points under $f$ is the set of critical values of $f$ and its complement is the set of regular values of $f$.

Now we can state the Smale theorem:

THEOREM 3.3. *If $E$ and $F$ are Banach spaces and $f: E \to F$ is a Fredholm map of class $C^q$, with $q > \max\{0, \operatorname{ind} f\}$, then the set of regular values of $f$ is residual (a countable intersection of open dense sets) in $F$. Moreover, if $y$ is a regular value of $f$, then $f^{-1}(y)$ is either empty or a manifold of dimension* ind $f$ (*if* ind $f = 0$, *then* $f^{-1}(y)$ *is discrete*).

REMARK 3.2. By Baire's category theorem, the set of regular values of $f$ is dense in $F$.

We now reconsider the system (1.13)−(1.17), defining $\mathbf{v} = \mathbf{u}/a$, $T = s + \tau_h$ and introducing an arbitrary quantity of heat $Q \in L^2(\Omega)$ in (1.14); the corresponding system in which we are interested is:

$$\operatorname{div} \mathbf{v} = 0 \quad \text{in } \Omega, \tag{3.21}$$

$$P_H(\mathbf{Bv} + T\mathbf{g}) = 0 \quad \text{in } \mathbf{L}^2(\Omega), \tag{3.22}$$

$$-\operatorname{div}(\mathbf{A}\nabla T) + a\mathbf{v}\nabla T = Q \quad \text{in } \Omega, \tag{3.23}$$

$$\mathbf{v}\cdot\boldsymbol{\nu} = 0 \quad \text{on } \partial\Omega, \tag{3.24}$$

$$T = \tau \quad \text{on } \partial\Omega \cdot \tag{3.25}$$

We denote by $\mathscr{S}(a, Q, \tau)$ the set of the weak solutions $(\mathbf{v}, t)$ of the problem (3.21)−(3.25); it can be proved that $\mathscr{S}(a, Q, \tau)$ is a non-empty compact subset of $\mathbf{H} \cap \mathbf{H}^2(\Omega) \times H^2(\Omega)$. We can use the same techniques as in Sections 1 and 2, and everything holds identically except the weak maximum principle (Theorem 1.3), which takes the following form:

THEOREM 3.4. *If* $(\mathbf{v}, T) \in \mathbf{H} \times H^1(\Omega)$ *is a solution of* (3.21)−(3.25)), *then* $T \in L^\infty(\Omega)$ *and*

$$|T|_\infty \leqslant \frac{1}{2} + C(\Omega)|Q|. \tag{3.26}$$

*Proof.* Passing through the Neumann problem of the pressure as in Lemma 1.1, we get $\mathbf{u} \in H^1(\Omega)$; thus we can take the duality product of (3.23) with any $S \in H_0^1(\Omega)$:

$$A(T, S) + ab(\mathbf{v}, T, S) = (Q, S). \tag{3.27}$$

For any $k > 0$ we define $S_k = \operatorname{sgn}(T) \max\{|T| - \frac{1}{2} - k, 0\}$; according to Proposition 5.2.2 in Chapter 1, $S_k \in H_0^1(\Omega)$ and, if

$$A_k = \left\{x \in \Omega \,\middle|\, |T| > \frac{1}{2} + k\right\},$$

we see that $\nabla S_k = \nabla T$ on $A_k$ and $\nabla S_k = 0$ elsewhere. Choosing $S = S_k$ in (3.27), we have

$$C_1|S_k|^2_{L^6(A_k)} \leqslant \|S_k\|^2 \leqslant A(S_k, S_k) = A(S_k, S_k) + ab(\mathbf{u}, S_k, S_k) =$$

$$= A(T, S_k) + ab(\mathbf{u}, T, S_k) = (Q, S_k) \leqslant |Q|(\text{meas } A_k)^{1/3}|S_k|_{L^6(A_k)},$$

that is

$$|S_k|_{L^6(A_k)} \leqslant C_2(\text{meas } A_k)^{1/3}|Q|. \tag{3.28}$$

Since, for any $h > k$,

$$(h-k)(\text{meas } A_h)^{1/6} \leqslant |S_k|_{L^6(A_h)} \leqslant |S_k|_{L^6(A_k)},$$

from (3.28) it follows that

$$\text{meas } A_h \leqslant C_3 \frac{|Q|^6}{(h-k)^6}(\text{meas } A_k)^2 \tag{3.29}$$

Because meas $A_k$ is a non-increasing function (after $k$), then by using a classical lemma (see Kinderlehrer and Stampacchia (1980), Chap. 2, Appendix B) we can show that meas $A_d = 0$ for $d = C_4\,(\text{meas}\,A_0)^{1/6}\,|Q|$. □

Defining now the Banach spaces

$$E_1 = \{(\mathbf{v}, T) \in \mathbf{H} \cap \mathbf{H}^2(\Omega) \times H^2(\Omega) \,|\, P_H(\mathbf{B}\mathbf{v} + T\mathbf{g}) = 0\}, \tag{3.30}$$

$$F = L^2(\Omega) \times H^{3/2}(\partial\Omega), \tag{3.31}$$

and the map $f_1 : E_1 \rightarrow F$ by

$$f_1(\mathbf{v}, T) = (-\operatorname{div}(\mathbf{A}\nabla T) + a\mathbf{v}\nabla T, T/\partial\Omega), \tag{3.32}$$

the system (3.21)−(3.25) is equivalent, in the sense of Propositions 1.1 and 1.2, to

$$f_1(\mathbf{v}, T) = (Q, \tau). \tag{3.33}$$

LEMMA 3.5. $f^1 : E_1 \rightarrow F$, *defined by* (3.30)−(3.32), *is a Fredholm map with* ind $f_1 = 0$.

*Proof.* The Fréchet differential of $f_1$, given by

$$\langle Df_1(\mathbf{v}, T), (\mathbf{u}, S)\rangle = (-\operatorname{div}(\mathbf{A}\nabla S) + a\mathbf{u}\nabla T + a\mathbf{v}\nabla S, S/\partial\Omega),$$

has the form $L + K$, where

$$L(\mathbf{u}, s) = (-\operatorname{div}(\mathbf{A}\nabla S), S/\partial\Omega)$$

is an isomorphism from $E_1$ onto $F$, and

$$K(\mathbf{u}, S) = a(\mathbf{u}\nabla T + \mathbf{v}\nabla S, 0/\partial\Omega)$$

is a compact operator from $E_1$ with values in $F$, because for any

$(\mathbf{v}, T) \in \mathbf{H} \cap \mathbf{H}^2(\Omega) \times H^2(\Omega)$, the operator

$$(\mathbf{u}, S) \rightarrow (\mathbf{u}\nabla T + \mathbf{v}\nabla S)$$

is continuous from $\mathbf{H} \cap \mathbf{H}^1(\Omega) \times H^1(\Omega)$ with values in $L^2(\Omega)$. Hence $Df_1(\mathbf{v}, T)$ is a Fredholm operator and

$$\operatorname{ind} Df_1(\mathbf{v}, T) = \operatorname{ind}(L + K) = \operatorname{ind} L = 0.$$ □

THEOREM 3.5. *For every* $a \geqslant 0$ *there exists a dense open set* $\mathcal{O}_1$ *in* $F$, *such that for any* $(Q, \tau) \in \mathcal{O}_1$ *the set* $\mathcal{S}(a, Q, \tau)$ *is finite. Moreover, if* $\mathcal{O}_\alpha$ *is a connected component of* $\mathcal{O}_1$, *then for any* $(Q, \tau) \in \mathcal{O}_\alpha$ *the number of elements of* $\mathcal{S}(a, Q, \tau)$ *is constant and every element of* $\mathcal{S}(a, Q, \tau)$ *is a* $C^\infty$ *function of the pair* $(Q, \tau)$.

*Proof.* Obviously $f_1$ is of class $C^\infty$ and, from Lemma 3.5, we find that we can use the hypothesis put forward in Theorem 3.3. Denoting by $\mathcal{O}_1$ the set of regular values of $f_1$, and bearing in mind that $\mathcal{S}(a, Q, \tau)$ is compact in $\mathbf{H} \cap \mathbf{H}^2(\Omega) \times H^2(\Omega)$ − and hence in $E_1$, everything can be proved exactly as in Foiaş and Temam (1977). □

We now pass to the second application of Smale's theorem. This time we put

$$E_2 = E_1 \times \mathbb{R}_+ \tag{3.34}$$

and define $f_2$: $E_2 \to F$ by

$$f_2(\mathbf{v}, T, a) = (-\operatorname{div}(\mathbf{A}\nabla T) + a\,\mathbf{v}\nabla T, T/\partial\Omega). \tag{3.35}$$

LEMMA 3.6. $f_2 : E_2 \to F$, *defined by* (3.34) *and* (3.35), *is a Fredholm map with* $\operatorname{ind} f_2 = 1$.

*Proof.* The Fréchet differential of $f_2$, given by

$$\langle Df_2(\mathbf{v}, T, a), (\mathbf{u}, S, b)\rangle = (-\operatorname{div}(\mathbf{A}\nabla S) + a\mathbf{v}\nabla S + a\mathbf{u}\nabla T + \\ + b\mathbf{v}\nabla T, S/\partial\Omega), \tag{3.36}$$

can be put in the form $L+K$, where

$$L(\mathbf{u}, S, b) = (-\operatorname{div}(\mathbf{A}\nabla S) + b\mathbf{v}\nabla T, S/\partial\Omega), \quad \forall b \in \mathbb{R}_+,$$

is an isomorphism from $E_1$ onto $F$, and

$$K(\mathbf{u}, S, b) = a(\mathbf{u}\nabla T + \mathbf{v}\nabla S, 0/\partial\Omega)$$

which is, as in Lemma 3.5, a compact operator. It follows that $Df_2(\mathbf{v}, T, a)$ is a Fredholm operator and that

$$\operatorname{ind} Df_2(\mathbf{v}, T, a) = \operatorname{ind}(L+K) = \operatorname{ind}(L) = 1. \qquad \square$$

THEOREM 3.6. *There exists a dense open set* $\mathcal{O}_2$ *in* $F$, *such that for any* $(Q, \tau) \in \mathcal{O}_2$ *the set*

$$\mathcal{S}(Q, \tau) = \{(\mathbf{v}, T, a) \in E_2 \mid (\mathbf{v}, T) \in \mathcal{S}(a, Q, \tau)\}$$

*is a one-dimensional manifold.*

*Proof.* The map $f_2$ is of class $C^\infty$; as an exercise the reader can check that

$$\langle (D^2 f(\mathbf{u}, R, a))(\mathbf{w}, T, c), (\mathbf{v}, S, b)\rangle = \\ = (c\mathbf{u}\nabla S + a\mathbf{w}\nabla S + b\mathbf{u}\nabla T + b\mathbf{w}\nabla R + a\mathbf{v}\nabla T + c\mathbf{v}\nabla R, 0/\partial\Omega).$$

As before, $\mathcal{O}_2$ is the set of regular values of $f_2$ and, as $\mathcal{S}(Q, \tau) = f_2^{-1}(Q, \tau)$, $\forall (Q, \tau) \in F$, almost the whole of Theorem 3.6 follows from Theorem 3.3; we have only to prove that $\mathcal{O}_2$ is open.

Let $(Q_n, \tau_n) \in F \setminus \mathcal{O}_2$ be a sequence converging in $F$ to $(Q, \tau)$. Let $(\mathbf{u}_n, T_n, 0) \in \mathcal{S}(Q_n, \tau_n)$; as $\{(Q_n, \tau_n)\}_n$ is bounded in $F$, from the corresponding estimations in Section 1 we have that $\{(\mathbf{u}_n, T_n)\}_n$ is bounded in $\mathbf{H} \cap \mathbf{H}^2(\Omega) \times H^2(\Omega)$. Hence, there exists $(\mathbf{u}, T) \in \mathbf{H} \cap \mathbf{H}^2(\Omega) \times H^2(\Omega)$ for which, passing, just in case, to a sub-sequence, we have

$$(\mathbf{u}_n, T_n) \rightharpoonup (\mathbf{u}, T) \text{ weakly in } \mathbf{H} \cap \mathbf{H}^2(\Omega) \times H^2(\Omega). \tag{3.37}$$

It follows at once that $(\mathbf{u}, T, 0) \in \mathcal{S}(Q, \tau)$. Moreover, $\operatorname{ind} f_2 = 1$, and

$$\dim\ker Df_2(\mathbf{u}_n, T_n, 0) = \dim\operatorname{coker} Df_2(\mathbf{u}_n, T_n, 0) + 1 \geqslant 1.$$

Thus, there exists $(\mathbf{v}_n, S_n, a) \in E_2$ with $(\mathbf{v}_n, S_n, a) \neq 0$ such that

$$\langle Df_2(\mathbf{u}_n, T_n, 0), (\mathbf{v}_n, S_m, a)\rangle = 0$$

which, via (3.36), is equivalent to

$$P_H(\mathbf{Bv} + S_n\,\mathbf{g}) = 0 \quad \text{in } \mathbf{L}^2(\Omega)\,, \tag{3.38}$$

$$-\operatorname{div}(\mathbf{A}\nabla S_n) + a_n\,\mathbf{u}_n\nabla T = 0 \quad \text{in } \Omega\,, \tag{3.39}$$

$$S_n = 0 \quad \text{on } \partial\Omega. \tag{3.40}$$

Note that $a_n = 0$ implies that $(\mathbf{v}_n\,,\, S_n) = 0$, and we can assume that $a_n = 1$, $(\mathbf{v}_n\,,\, S_n\,,\, a_n)$ being defined up to multiplication by a constant. From the system (3.38)–(3.40) with $a_n = 1$, it follows that $\{(\mathbf{v}_n, S_n)\}_n$ is bounded in $\mathbf{H} \cap \mathbf{H}^2(\Omega) \times H^2(\Omega)$, as the given term $\mathbf{u}_n \nabla T_n$ is bounded in $L^2(\Omega)$. Hence we can extract a new sub-sequence (still denoted by $n$) such that

$$(\mathbf{v}_n\,, S_n) \rightharpoonup (\mathbf{v}, S) \quad \text{weakly in } \mathbf{H} \cap \mathbf{H}^2(\Omega) \times H^2(\Omega)\,. \tag{3.41}$$

Now passing to the limit the system (3.38)–(3.40), from (3.37) and (3.41) we get

$$\langle Df_2(\mathbf{u}, T, 0), (\mathbf{v}, S, 1)\rangle = 0$$

with $(\mathbf{v}, S, 1) \in E_2$; as $(\mathbf{v}, S, 1) \neq 0$ the element $(\mathbf{u}, T, 0)$ is not a regular point, and hence $(Q, \tau) \in F \backslash \mathscr{O}_2$. □

REMARK 3.3. From Theorems 3.5 and 3.6 it follows that (*generical*) there are no bifurcation points and that fluctuations can appear only where the projection of $\mathscr{S}(Q, \tau)$ onto the Rayleigh number axis overlaps; also, as long as the Rayleigh number remains bounded, the number of overlaps is finite.

These results are closely related to the problem of natural convection considered previously because, as we have seen from Theorems 3.5 and 3.6, $Q$ can approximate the null value as much as we want.

## 4. The Evolution Case

We now examine the fundamental aspects of the time-dependent Darcy-Boussinesq system. The existence result is based on the construction of an approximate solution by the Galerkin method. Some a priori estimates are derived, the most important of them being a maximum principle, similar to that obtained for the steady case. Finally we prove a uniqueness theorem and a stability criterion.

Let $\Omega$ be an open, connected bounded set in $\mathbb{R}^n$ ($n = 2, 3$) locally located on one side of the boundary $\partial\Omega$, which is a manifold of class $C^2$ composed of a finite number of connected components; let $\theta > 0$ and let $Q$ denote the cylinder $\Omega \times (0, \theta)$. With the assumptions and approximations which are frequently used for thermal convection in a homogeneous porous medium saturated with an incompressible fluid, the governing evolution system of the Darcy – Boussinesq equations for the filtration velocity $\mathbf{u}$, the pressure $p$ and the temperature $T$ may be written as

$$\operatorname{div} \mathbf{u} = 0 \quad \text{in } Q, \tag{4.1}$$

$$\mathbf{B}\mathbf{u} + \nabla p = [1 - \alpha (T - T_r)]\, \mathbf{g} \quad \text{in } Q, \tag{4.2}$$

$$\sigma \frac{\partial T}{\partial t} + \mathbf{u} \nabla T = \operatorname{div} (\mathbf{A} \nabla T) \text{ in } Q. \tag{4.3}$$

The boundary and initial conditions are

$$\mathbf{u} \cdot \mathbf{v} = 0 \quad \text{on } \partial\Omega \times (0, \theta), \tag{4.4}$$

$$T = \tau \quad \text{on } \partial\Omega \times (0, \theta), \tag{4.5}$$

$$T(0) = T^0 \quad \text{in } \Omega, \tag{4.6}$$

where $\mathbf{v}$ is the unit outward normal, $\sigma > 0$ is a dimensionless number derived from the properties of the specific heat at constant volume of the porous medium, $\alpha > 0$, $T_r > 0$ and the positive symmetric tensors $\mathbf{A}$ and $\mathbf{B}$ have the same meaning as in the steady case. The other parameters in the problem, the gravitational acceleration $\mathbf{g}$, the temperature of the boundary $\tau$ and the initial condition $T^0$, are time-dependent, and are considered at a later stage (see Equations (4.7)−(4.9)).

For any function $f(x, t)$, of a space variable $x$ and a time variable $t$, we denote by $f(t)$ for any $t$ the mapping $x \mapsto f(x, t)$; then the function $t \mapsto f(t)$ is called the vector-valued function corresponding to $f(x, t)$. As from now on we have to handle such functions, we introduce their usual spaces, the notion of derivative and some basic properties, following Temam (1977, Chap. III).

Let $-\infty \leqslant a < b \leqslant +\infty$ and let $X$ be a Banach space normed by $|\cdot|_X$. For any $p \geqslant 1$ we say that a function $f\colon (a, b) \to X$ is $L^p$−strongly measurable if

$$t \to |f(t)|_X^p \quad \text{is measurable, for } p < \infty,$$

$$\operatorname{esssup}_{t \in (a, b)} |f(t)|_X < \infty, \quad p = \infty.$$

Then $L^p(a, b; X)$ denotes the space of $L^p$−strongly measurable functions defined on $(a, b)$ with values in $X$, which is a Banach space with the norm

$$|f|_{L^p(a, b; X)} = \left( \int_a^b |f(t)|_X^p \, \mathrm{d}t \right)^{1/p}, \quad p < \infty,$$

$$|f|_{L^\infty(a, b; X)} = \operatorname{esssup}_{t \in (a, b)} |f(t)|_X .$$

When $a$ and $b$ are finite we denote by $C^0(a, b; X)$ the space of continuous functions $f\colon [a, b] \to X$ equipped with the Banach norm

$$|f|_{C^0(a, b; X)} = \sup_{t \in [a, b]} |f(t)|_X .$$

PROPOSITION 4.1. *Let $X'$ be the dual of the Banach space $X$ and let $f, g \in L^1(a, b; X)$. Then, the following three conditions are equivalent*:

(a) *for some $\xi$ in $X$*,

$$f(t) = \xi + \int_a^t g(s)\, \mathrm{d}s \quad \text{a.e. } in\ (a, b),$$

(b) *for any* $\varphi$ *in* $\mathscr{D}((a, b))$,

$$\int_a^b f(t)\,\varphi'(t)\,\mathrm{d}t = -\int_a^b g(t)\,\varphi(t)\,\mathrm{d}t,$$

(c) *for any* $\eta$ *in* $X'$,

$$\frac{\mathrm{d}}{\mathrm{d}t}\langle f, \eta\rangle_X = \langle g, \eta\rangle_X \quad \text{in } \mathscr{D}'((a, b))$$

*where* $\langle , \rangle_X$ *denotes the duality product between* $X$ *and* $X'$. *Moreover, in each of these three equivalent cases,* $f$ *is almost everywhere equal to a function of* $C^0(a, b; X)$.

DEFINITION 4.1. The function $g$ in Proposition 4.1 is called the weak generalized derivative of $f$, and is denoted by

$$g = f'\left(= \frac{\mathrm{d}f}{\mathrm{d}t}\right).$$

Obviously, this definition can be extended to higher-order derivatives. Correspondingly we define the spaces $C^k(a, b; X)$, $k \geqslant 1$.

Coming back to the Darcy–Boussinesq system, we can now write the hypotheses for $\mathbf{g}$, $\tau$ and $T^0$

$$\mathbf{g} \in C^1(0, \theta; \mathbf{H}^2(\Omega)), \tag{4.7}$$

$$\tau \in C^2(0, \theta; H^{3/2}(\partial\Omega)), \tag{4.8}$$

$$T^0 \in H^2(\Omega), \qquad T^0 = \tau(0) \quad \text{on } \partial\Omega. \tag{4.9}$$

REMARK 4.1. The notation for the Sobolev spaces are the same as in the previous sections of this chapter. □

We now consider a system with homogeneous boundary conditions, using the following lemma:

LEMMA 4.1. *For any* $h > 0$ *there exists an element* $\tau_h \in C^2(0, \theta; H^2(\Omega))$ *satisfying almost everywhere in* $(0, \theta)$ *the conditions*

$$\tau_h = \tau - T_r \quad \text{on } \partial\Omega, \tag{4.10}$$

$$|S\nabla\tau_h| \leqslant h\,\|S\|, \quad \forall S \in H_0^1(\Omega). \tag{4.11}$$

*Proof.* The construction of $\tau_h$ is the same as in the proof of Lemma 5.1.1 in Chapter 1. The only new point is that the vector-valued function

$$L(\tau)(t) := L(\tau(t)),$$

where $L(\tau(t)) \in H^2(\Omega)$ is the element obtained by lifting the trace of $\tau(t) \in H^{3/2}(\partial\Omega)$, has the property

$$L(\tau) \in C^2(0, \theta; H^2(\Omega)).$$

This is a simple consequence of (4.8), and of the fact that the lifting operator

$$L: H^{3/2}(\partial\Omega) \to H^2(\Omega)$$

is linear and continuous. □

Later we shall choose a suitable value for the parameter $h > 0$; now we introduce

$$S = T - (\tau_h + T_r). \tag{4.12}$$

Thus the system (4.1)–(4.6) becomes

$$\operatorname{div} \mathbf{u} = 0 \quad \text{in } Q, \tag{4.13}$$

$$\mathbf{Bu} + \nabla p = [1 - \alpha(S + \tau_h)]\,\mathbf{g} \quad \text{in } Q, \tag{4.14}$$

$$\sigma \frac{\partial S}{\partial t} + \mathbf{u}\nabla(S + \tau_h) = \operatorname{div}(\mathbf{A}\nabla(S + \tau_h)) - \sigma \frac{\partial \tau_h}{\partial t} \quad \text{in } Q, \tag{4.15}$$

$$\mathbf{u}\cdot\mathbf{v} = 0 \quad \text{on } \partial\Omega \times (0, \theta), \tag{4.16}$$

$$S = 0 \quad \text{on } \partial\Omega \times (0, \theta), \tag{4.17}$$

$$S(0) = S^0 \quad \text{in } \Omega, \tag{4.18}$$

where $S^0 = T^0 - (\tau_h(0) + T_r) \in H^2(\Omega) \cap H_0^1(\Omega)$.

Let us assume that $\mathbf{u}$, $p$ and $S$ are smooth solutions of (4.13)–(4.18); then, taking the dual product of (4.14) and (4.15) with

$$\mathbf{v} \in \mathscr{V}(\Omega) = \{\mathbf{v} \in \mathscr{D}(\Omega) \mid \operatorname{div} \mathbf{v} = 0 \text{ in } \Omega\},$$

and similarly for $T \in \mathscr{D}(\Omega)$, one can easily check that

$$(\mathbf{Bu}, \mathbf{v}) + \alpha(S\mathbf{g}, \mathbf{v}) = ((1 - \alpha\tau_h)\,\mathbf{g}, \mathbf{v}), \tag{4.19}$$

$$\sigma \frac{\mathrm{d}}{\mathrm{d}t}(S, T) + (\mathbf{A}\nabla S, \nabla T) + (\mathbf{u}\nabla(S + \tau_h), T) =$$
$$= (\operatorname{div}(\mathbf{A}\nabla\tau_h) - \sigma\tau_h', T). \tag{4.20}$$

As $\mathscr{V}(\Omega)$ and $\mathscr{D}(\Omega)$ are dense in $\mathbf{H}$ and $H_0^1(\Omega)$ respectively, the relations (4.19) and (4.20) suggest the following variational formulation of the problem (4.13)–(4.18):

PROBLEM 4.1. Find $\mathbf{u} \in L^\infty(0, \theta; \mathbf{H}) \cap L^2(0, \theta; \mathbf{H}^1(\Omega))$ and $S \in L^2(0, \theta; H_0^1(\Omega)) \cap L^\infty(0, \theta; L^2(\Omega))$ satisfying the initial condition (4.18) and the system given by (4.19) and (4.20), for any $(\mathbf{v}, T) \in \mathbf{H} \times H_0^1(\Omega)$.

The initial condition (4.18) does not seem to make sense. Therefore, let us consider the functions

$$t \mapsto A(S)(t) \in H^{-1}(\Omega),$$

$$t \mapsto B(\mathbf{u}, S)(t) \in H^{-1}(\Omega),$$

defined a.e. in $(0, \theta)$ for any $T \in H_0^1(\Omega)$ by

$$\langle A(S), T\rangle = (\mathbf{A}\nabla S, \nabla T), \tag{4.21}$$

$$\langle B(\mathbf{u}, S), T\rangle = (\mathbf{u}\,\nabla S, T) = -(\mathbf{u}\,\nabla T, S), \tag{4.22}$$

where $(\mathbf{u}, S)$ is a solution of Problem 4.1 and $\langle\,,\rangle$ denotes the duality product between $H^{-1}(\Omega)$ and $H_0^1(\Omega)$. Then, introducing

$$\mathrm{F} := \operatorname{div}(\mathbf{A}\nabla\tau_h) - \sigma\tau_h', \tag{4.23}$$

$$\mathbf{G} := (1 - \alpha\tau_h)\,\mathbf{g}, \tag{4.24}$$

we can write (4.20) as

$$\sigma\frac{\mathrm{d}}{\mathrm{d}t}(S, T) = \langle F - A(S) - B(\mathbf{u}, S + \tau_h), T\rangle, \quad \forall T \in H_0^1(\Omega). \tag{4.25}$$

REMARK 4.2. The hypotheses given by (4.7) and (4.8) imply directly that $F \in C^1(0, \theta; L^2(\Omega))$ and $\mathbf{G} \in C^1(0, \theta; \mathbf{H}^2(\Omega))$.

If $A_\infty$ is the last eigenvalue of $\mathbf{A}$, then

$$|A(S)|_{L^2(0,\theta;H^{-1}(\Omega))} \leqslant \left(\int_0^\theta |\mathbf{A}\nabla S(t)|^2\,\mathrm{d}t\right)^{1/2} \leqslant A_\infty |S|_{L^2(0,\theta;H_0^1(\Omega))} \tag{4.26}$$

Also, by the relevant Sobolev inequalities we have

$$|B(\mathbf{u}, S)|_{L^1(0,\theta;H^{-1}(\Omega))} \leqslant \int_0^\theta |\mathbf{u}(t)|_4\,|S(t)|_4\,\mathrm{d}t \leqslant$$

$$\leqslant C|\mathbf{u}|_{L^2(0,\theta;\mathbf{H}^1(\Omega))}\,|S|_{L^2(0,\theta;H_0^1(\Omega))}\,. \tag{4.27}$$

The estimations (4.26) and (4.27) yield

$$[F - A(S) - B(\mathbf{u}, S + \tau_h)] \in L^1(0, \theta; H^{-1}(\Omega))\,.$$

Hence Proposition 4.1, applied to (4.25), implies that $S' \in L^1(0, \theta; H^{-1}(\Omega))$, and

$$\sigma S' = F - A(S) - B(\mathbf{u}, S + \tau_h)\,. \tag{4.28}$$

Moreover, $S$ is a.e. equal to a function of $C^0(0, \theta; H^{-1}(\Omega))$, and thus the initial condition (4.18) is seen to make sense.

In order to prove the existence of the generalized solutions of the Problem 4.1, we employ the Galerkin method. The starting point is that $\mathscr{V}(\Omega)$ and $\mathscr{D}(\Omega)$ are dense in the separable spaces $\mathbf{H}$ and $H_0^1(\Omega)$ respectively, and hence there exist the sequences

$$\{\mathbf{v}_j\}_{j\in\mathbb{N}} \subseteq \mathscr{V}(\Omega), \qquad \{T_j\}_{j\in\mathbb{N}} \subseteq \mathscr{D}(\Omega), \tag{4.29}$$

such that they are linearly independent and the finite combinations are dense in $\mathbf{H}$ and $\mathbf{H}_0^1(\Omega)$, respectively:

Denoting by $\mathbf{H}_i$ the space spanned by the set $\{\mathbf{v}_j \mid j = 1, \ldots, i\}$ and by $W_i$ the space spanned by the set $\{T_j \mid j = 1, \ldots, i\}$, we define, for each $i \in \mathbb{N}$, an approximate solution $(\mathbf{u}_i, S_i)$ of $(\mathbf{u}, S)$ by the following problem:

PROBLEM 4.2. Find $(\mathbf{u}_i, S_i)$ of the form

$$\mathbf{u}_i(t) = \sum_{j=1}^{i} u_{ij}(t)\,\mathbf{v}_j, \qquad S_i(t) = \sum_{j=1}^{i} S_{ij}(t)\,T_j \tag{4.30}$$

satisfying for any $k \in \{1, \dots, i\}$ the system

$$(\mathbf{B}\mathbf{u}_i, \mathbf{v}_k) + \alpha\,(S_i\,\mathbf{g}, \mathbf{v}_k) = (G, \mathbf{v}_k), \tag{4.31}$$

$$\sigma\,(S'_i, T_k) + (\mathbf{A}\nabla S_i, \nabla T_k) + (\mathbf{u}_i\nabla(S_i + \tau_h), T_k) = (F, T_k), \tag{4.32}$$

$$S_i(0) = S_i^0 \in W_i, \quad \text{such that } S_i^0 \to S^0 \text{ strongly in } H_0^1(\Omega). \tag{4.33}$$

Our object now is to show that, for each $i$, Problem 4.2 has at least one solution. First we write (4.31) – (4.33) in terms of its effective unknows $u_{ij}(t)$ and $S_{ij}(t)$:

$$\sum_{j=1}^{i} u_{ij}(t)\,(\mathbf{B}\mathbf{v}_j, \mathbf{v}_k) + \alpha \sum_{l=1}^{i} S_{il}(t)\,(T_l\,\mathbf{g}(t), \mathbf{v}_k) = (\mathbf{G}(t), \mathbf{v}_k), \tag{4.34}$$

$$\sigma \sum_{j=1}^{i} (T_j, T_k)\,\frac{\mathrm{d}S_{ij}}{\mathrm{d}t}(t) + \sum_{j=1}^{i} S_{ij}(t)\,(\mathbf{A}\nabla T_j, \nabla T_k) + \sum_{j=1}^{i} u_{ij}(t)\,(\mathbf{v}_j\nabla\tau_h(t), T_k) +$$

$$+ \sum_{j=1}^{i}\sum_{l=1}^{i} u_{ij}(t)\,S_{il}(t)\,(\mathbf{v}_j\nabla T_l, T_k) = (F(t), T_k), \tag{4.35}$$

$$S_{ik}(0) = S_{ik}^0 := \text{the } k\text{th component in } W_i \text{ of } S_i^0,\ k \in \{1, \dots, i\}. \tag{4.36}$$

As the matrix $(\mathbf{B}\mathbf{v}_j, \mathbf{v}_k)$ is non-singular because $\mathbf{B}$ is positive-definite and $\mathbf{v}_1, \dots, \mathbf{v}_i$ are linearly independent, and denoting by $\mathbf{D}$ the inverse of this $i \times i$ matrix, from (4.34) we get

$$u_{ij}(t) = -\,\alpha \sum_{k=1}^{i}\sum_{l=1}^{i} D_{kj}\,S_{il}(t)\,(T_l\,\mathbf{g}, \mathbf{v}_k) + \sum_{k=1}^{i} D_{kj}\,(\mathbf{G}(t), \mathbf{v}_k). \tag{4.37}$$

Using (4.37), and inverting the non-singular matrix $[(T_j, T_k)]$, (4.35) becomes a system of the form

$$\frac{\mathrm{d}S_{ij}}{\mathrm{d}t}(t) = \varphi_j\,(t; S_{i1}(t), \dots, S_{ii}(t)), \quad j = 1, \dots, i, \tag{4.38}$$

with $\varphi_j$ at least continuous. The ordinary differential system (4.38) with the initial condition (4.36) then has a local maximal solution defined on some interval $[0, \theta_i)$. If $\theta_i < \theta$, then necessarily $\| S_i(t) \| \to \infty$ as $t \to \theta_i^-$. Therefore, if we can show that $\|S_i(t)\|$ is bounded independently of $i$ and $t$, we will have proved that $\theta_i = \theta$ for any $i$, and that $S_i(t)$ – and, implicitly, $\mathbf{u}_i(t)$ – are in fact the global solutions of the Problem 4.2. As usual, this will be proved by some a priori estimates.

Let $t \in [0, \theta_i)$; multiplying (4.34) by $u_{ik}(t)$ and adding these equations for $k = 1, \dots, i$, we obtain

$$(\mathbf{B}\mathbf{u}_i(t), \mathbf{u}_i(t)) + \alpha\,(S_i(t)\,\mathbf{g}, \mathbf{u}_i(t)) = (\mathbf{G}(t), \mathbf{u}_i(t)). \tag{4.39}$$

Denoting by $B_1$ the first eigenvalue of $\mathbf{B}$, we have from (4.39) the important relation

$$|\mathbf{u}_i(t)| \leqslant (\alpha g|S_i(t)| + |\mathbf{G}(t)|)/B_1\,, \tag{4.40}$$

where $g = |\mathbf{g}|_{L^\infty(0,\,\theta;\,L^\infty(\Omega))}$. Next, we multiply (4.35) by $S_{ik}(t)$ and sum over $k$ from 1 to $i$. Taking in account that $(\mathbf{u}_i \nabla S_i\,,\, S_i) = 0$, we get

$$\sigma\,(S'_i(t), S_i(t)) + (\mathbf{A}\nabla S_i(t), \nabla S_i(t)) +$$

$$+ (\mathbf{u}_i(t)\,\nabla\tau_h(t), S_i(t)) = (F(t), S_i(t))\,. \tag{4.41}$$

Denoting by $A_1$ the first eigenvalue of $\mathbf{A}$, from (4.41) we obtain

$$\sigma\frac{\mathrm{d}}{\mathrm{d}t}|S_i(t)|^2 + 2A_1\|S_i(t)\|^2 \leqslant 2\,(F(t), S_i(t)) - 2\,(\mathbf{u}_i(t)\nabla\tau_h(t), S_i(t)) \leqslant$$

$$\leqslant 2|F(t)||S_i(t)| + 2\,|\mathbf{u}_i(t)||S_i(t)\,\nabla\tau_h(t)|\,. \tag{4.42}$$

Introducing in (4.42) the relation (4.40), and using the property (4.11) of $\tau_h$, it follows that

$$\sigma\frac{\mathrm{d}}{\mathrm{d}t}|S_i(t)|^2 + 2A_1\|\dot{S}_i(t)\|^2 \leqslant 2|F(t)||S_i(t)| +$$

$$+ 2h\|S_i(t)\|(\alpha g|S_i(t)| + |\mathbf{G}(t)|)/B_1\,. \tag{4.43}$$

Recalling the Friedrichs inequality (see (1.25))

$$|S| \leqslant 2L\|S\|/a_0, \quad \forall S \in H_0^1(\Omega)\,, \tag{4.44}$$

where $L$ is the length of the $n$-cube in which $\Omega$ can be included, then from the inequality

$$2ab \leqslant a^2/\varepsilon + \varepsilon b^2, \quad a, b \in \mathbb{R}, \varepsilon > 0\,, \tag{4.45}$$

and from (4.43) we have

$$\sigma\frac{\mathrm{d}}{\mathrm{d}t}|S_i(t)|^2 + (3A_1/2 - h^2 - 4\alpha ghL/B_1 a_0)\|S_i(t)\|^2 \leqslant$$

$$\leqslant 8L^2|F(t)|^2/A_1 a_0^2 + |\mathbf{G}(t)|^2/B_1^2\,. \tag{4.46}$$

With $h$ sufficiently small, such that

$$h^2 + 4\alpha ghL/B_1 a_0 < A_1/2\,, \tag{4.47}$$

it follows that

$$\sigma\frac{\mathrm{d}}{\mathrm{d}t}|S_i(t)|^2 + A_1\|S_i(t)\|^2 < C_1\,, \tag{4.48}$$

where $C_1 = 8L^2|F|^2_{C^0(0,\,\theta;\,L^2(\Omega))}/A_1 a_0^2 + |\mathbf{G}|^2_{C^0(0,\,\theta;\,\mathbf{L}^2(\Omega))}/B_1^2\,.$

Integrating (4.48) from 0 to some $s \in [0, \theta]$, we obtain

$$\sigma|S_i(s)|^2 + A_1\int_0^s \|S_i(t)\|^2\,\mathrm{d}t < \sigma|S_i(0)|^2 + C_1 s \leqslant \sigma|S^0|^2 + C_1\theta. \tag{4.49}$$

In particular, from (4.49) we find that

the sequence $\{S_i\}$ is bounded in $L^\infty(0, \theta; L^2(\Omega))$. (4.50)

With this result, and recalling that $\mathbf{G} \in L^\infty(Q)$ from the estimation (4.40), we get

the sequence $\{\mathbf{u}_i\}$ is bounded in $L^\infty(0, \theta; \mathbf{H})$. (4.51)

Another important result is obtained by setting $s = \theta$ in (4.49):

the sequence $\{S_i\}$ is bounded in $L^2(0, \theta; H_0^1(\Omega))$. (4.52)

From the Dunford – Pettis theorem (see for instance Yosida (1965)) we learn that $L^\infty(0, \theta; L^2(\Omega))$ is the dual of $L^1(0, \theta; L^2(\Omega))$. Thus (4.50) shows that there exists an element $S \in L^\infty(0, \theta; L^2(\Omega))$ for which, on some sub-sequence, we have:

$S_i \rightharpoonup S$ in the weak-star topology of $L^\infty(0, \theta; L^2(\Omega))$. (4.53)

That is, for any $T \in L^1(0, \theta; L^2(\Omega))$

$$\int_0^\theta \int_\Omega (S_i - S)\, T \mathrm{d}\Omega\, \mathrm{d}t \to 0\,. \tag{4.54}$$

From (4.52) we see that there exists an element $S^* \in L^2(0, \theta; H_0^1(\Omega))$ for which, on some sub-sequence still denoted by $S_i$, we get

$S_i \rightharpoonup S^*$ weakly in $L^2(0, \theta; H_0^1(\Omega))$. (4.55)

That is, for any $T \in L^2(0, \theta; H^{-1}(\Omega))$,

$$\int_0^\theta < T, S_i - S^* > \mathrm{d}t \to 0\,. \tag{4.56}$$

Since $<\, ,\, >$ is an extension of the scalar product in $L^2(\Omega)$ (because of the identification of $[L^2(\Omega)]'$ with $L^2(\Omega)$), it follows from (4.54) and (4.56) that $S = S^*$.

From (4.51) we obtain some $\mathbf{u} \in L^\infty(0, \theta; \mathbf{H})$ for which, with another passage to a sub-sequence, we have

$\mathbf{u}_i \rightharpoonup \mathbf{u}$ weakly star in $L^\infty(0, \theta; \mathbf{H})$. (4.57)

That is, for any $\mathbf{v} \in L^1(0, \theta; \mathbf{H})$,

$$\int_0^\theta \int_\Omega (\mathbf{u}_i - \mathbf{u})\, \mathbf{v}\, \mathrm{d}\Omega\, \mathrm{d}t \to 0\,. \tag{4.58}$$

Now, we fix some $k \in \mathbb{N}$ and pass with $i$ to the infinite in Equation (4.31). It is easily shown that

$$(\mathbf{B}\mathbf{u}, \mathbf{v}_k) + \alpha\, (S\mathbf{g}, \mathbf{v}_k) = (\mathbf{G}, \mathbf{v}_k)$$

and, since every term depends linearly and continuously (in $\mathbf{H}$) on the $\mathbf{v}_k$-argument, the implication is that $(\mathbf{u}, S)$ satisfies (4.19) for any $\mathbf{v} \in \mathbf{H}$.

As in Proposition 1.2, we can show that $\exists\, p \in L^\infty(0, \theta; H^1(\Omega))$ such that (4.14) is satisfied in $L^\infty(0, \theta; \mathbf{L}^2(\Omega))$. Taking into account (4.13) and (4.16) – in fact, the definition of $\mathbf{H}$ – it follows that $p$ has to satisfy the conditions

$$\operatorname{div}(\mathbf{K}\nabla p) = \operatorname{div}[\mathbf{K}(1-\alpha(S+\tau_h))\mathbf{g}] \quad \text{in } \Omega \text{ a.e. in } (0,\theta), \tag{4.59}$$

$$(\mathbf{K}\nabla p)\cdot\mathbf{v} = [\mathbf{K}(1-\alpha(\tau - T_r))\mathbf{g}]\,\mathbf{v} \quad \text{on } \partial\Omega \text{ a.e. in } (0,\theta). \tag{4.60}$$

As $\operatorname{div}[\mathbf{K}(1-\alpha(S+\tau_h))\mathbf{g}] \in L^2(0,\theta;L^2(\Omega))$ and $\mathbf{K}(1-\alpha(\tau - T_r))\mathbf{g}\cdot\mathbf{v} \in C^1(0,\theta;H^{3/2}(\partial\Omega)) \subseteq L^2(0,\theta;H^{1/2}(\partial\Omega))$, by the regularity theorem for the Neumann problem (see for instance Lions and Magenes (1968), Chap. 2 we obtain $p \in L^2(0,\theta;H^2(\Omega))$. Recalling (4.14), we obtain immediately

$$\mathbf{u} \in L^2(0,\theta;\mathbf{H}^1(\Omega)). \tag{4.61}$$

In order to pass to the limit the equation (4.32), we multiply it by a scalar function $\psi(t)$ of class $C^1([0,\theta])$ such that $\psi(\theta)=0$. After integration on $[0,\theta]$, we obtain

$$\int_0^\theta (\mathbf{u}_i\nabla(S_i+\tau_h), T_k\psi)\,\mathrm{d}t = \int_0^\theta (F, T_k\psi)\,\mathrm{d}t - \int_0^\theta (\mathbf{A}\nabla S_i, \nabla T_k)\,\psi\,\mathrm{d}t +$$

$$+\,\sigma\int_0^\theta (S_i, T_k)\,\psi'\,\mathrm{d}t + \sigma\psi(0)\,(S_i^0, T_k)\,. \tag{4.62}$$

By using (4.33) and (4.55), the right-hand side of (4.62) is shown to converge to

$$\int_0^\theta (F, T_k\psi)\,\mathrm{d}t - \int_0^\theta (\mathbf{A}\nabla S, \nabla T_k)\,\psi\,\mathrm{d}t + \sigma\int_0^\theta (S, T_k)\,\psi'\,\mathrm{d}t + \sigma\psi(0)\,(S^0, T_k).$$

As the injection of $H_0^1(\Omega)$ in $L^2(\Omega)$ is compact, (4.55) also yields

$$S_i \to S \quad \text{strongly in } L^2(0,\theta;L^2(\Omega)). \tag{4.63}$$

Denoting by $R_i := \mathbf{u}_i\psi\,\nabla T_k \in L^\infty(0,\theta;L^2(\Omega))$, we see that

$$\int_0^\theta (\mathbf{u}_i\nabla(S_i+\tau_h), T_k\psi)\,\mathrm{d}t = -\int_0^\theta (R_i, S_i+\tau_h)\,\mathrm{d}t$$

and, as in particular

$$R_i \rightharpoonup R = \mathbf{u}\psi\,\nabla T_k \quad \text{weakly in } L^2(0,\theta;L^2(\Omega)),$$

we obtain

$$\int_0^\theta (\mathbf{u}_i\,\nabla(S_i+\tau_h), T_k\psi)\,\mathrm{d}t \to -\int_0^\theta (R, S+\tau_h)\,\mathrm{d}t =$$

$$= \int_0^\theta (\mathbf{u}\,\nabla(S+\tau_h), T_k\psi)\,\mathrm{d}t\,.$$

Hence, from (4.62) we obtain in the limit,

$$\int_0^\theta (\mathbf{u}\nabla(S+\tau_h), T_k)\psi\, dt = \int_0^\theta (F, T_k)\psi\, dt - \int_0^\theta (\mathbf{A}\nabla S, \nabla T_k)\psi\, dt +$$

$$+\sigma\int_0^\theta (S, T_k)\psi'\, dt + \sigma\psi(0)(S^0, T_k). \tag{4.64}$$

The regularity property (4.61) permit us to evaluate the nonlinear term

$$|(\mathbf{u}\nabla(S+\tau_h), T_k)| = |(\mathbf{u}\nabla T_k, S+\tau_h)| \leqslant |\mathbf{u}|_4 \|T_k\| \, |S+\tau_h|_4 \leqslant$$

$$\leqslant C\|\mathbf{u}\|_1 \|S+\tau_h\|_1 \, \|T_k\|,$$

which is thus shown to be continuous (in $H_0^1$) on the $T_k$-argument. As all the other terms in (4.64) obviously share this property, we obtain for any $T \in H_0^1(\Omega)$

$$\int_0^\theta (\mathbf{u}\nabla(S+\tau_h), T)\psi\, dt = \int_0^\theta (F, T)\psi\, dt + \sigma\psi(0)(S^0, T) +$$

$$+\sigma\int_0^\theta (S, T)\psi'\, dt - \int_0^\theta (\mathbf{A}\nabla S, \nabla T)\psi\, dt. \tag{4.65}$$

Taking $\psi \in \mathscr{D}((0, \theta))$, we observe that (4.65) is exactly (4.20) in the distribution sense. Therefore, in order to prove that $(\mathbf{u}, S)$ is a solution of Problem 4.1, it remains to show that $S$ satisfies the initial condition (4.18). For this we take $\psi \in C^1([0, \theta])$ with $\psi(\theta) = 0$ and $\psi(0) \neq 0$, and multiply (4.20) by it. After integration on $[0, \theta]$, we obtain

$$\int_0^\theta \mathbf{u}\nabla(S+\tau_h), T)\psi\, dt = \int_0^\theta (F, T)\psi\, dt + \sigma\psi(0)(S(0), T) +$$

$$+\sigma\int_0^\theta (S, T)\psi'\, dt - \int_0^\theta (\mathbf{A}\nabla S, \nabla T)\psi\, dt. \tag{4.66}$$

Subtracting (4.65) from (4.66) gives

$$(S(0) - S^0, T) = 0, \quad \forall\, T \in H_0^1(\Omega),$$

which obviously implies (4.18).

Thus we have proved:

THEOREM 4.1. *Problem* 4.1 *has at least one solution.*

After the following weak-maximum principle, we shall study the uniqueness.

THEOREM 4.2. *If* $(\mathbf{u}, S)$ *is a solution of Problem* 4.1, *then* $S \in L^\infty(0, \theta; L^\infty(\Omega))$ *with*

$$|S+\tau_h|_{L^\infty(0,\theta;L^\infty(\Omega))} \leqslant C(T^0), \tag{4.67}$$

*where* $C(T^0) = \max\{|\tau - T_r|_{L^\infty(0,\theta;L^\infty(\partial\Omega))}, |T^0 - T_r|_\infty\}$.

*Proof.* As $S \in L^2(0, \theta; H_0^1(\Omega))$, the property (4.10) implies that

$$S(t) + \tau_h(t) = \tau(t) - T_r \quad \text{on } \partial\Omega \text{ a.e. in } (0, \theta).$$

Then, from Proposition 5.2.2 in Chapter 1, we have

$$R(t) = \max\{S(t) + \tau_h(t) - C(T^0), 0\} \; L^2(0, \theta; H_0^1(\Omega)).$$

From Proposition 5.2.3 in Chapter 1, we have

$$\nabla R = \begin{cases} \nabla(S + \tau_h), & R \neq 0, \\ \mathbf{0} & R = 0, \end{cases} \tag{4.68}$$

$$R' = \begin{cases} S' + \tau'_h, & R \neq 0, \\ 0 & R = 0. \end{cases} \tag{4.69}$$

Consequently,

$$\frac{\mathrm{d}}{\mathrm{d}t} |R|^2 = 2(R', R) = 2(S' + \tau'_h, R). \tag{4.70}$$

Then, setting $T = R$ in (4.20) and taking into account (4.68)–(4.70), we get

$$0 = \frac{\sigma}{2}\frac{\mathrm{d}}{\mathrm{d}t}|R|^2 + (\mathbf{A}\nabla R, \nabla R) + (\mathbf{u}\,\nabla R, R) = \frac{\sigma}{2}\frac{\mathrm{d}}{\mathrm{d}t}|R^2 + (\mathbf{A}\nabla R, \nabla R) \geqslant$$

$$\geqslant \frac{\sigma}{2}\frac{\mathrm{d}}{\mathrm{d}t}|R|^2 + A_1\,\|R\|^2. \tag{4.71}$$

In particular, we have

$$\frac{\mathrm{d}}{\mathrm{d}t}|R|^2 \leqslant 0,$$

and by integration on $(0, t)$

$$|R(t)|^2 \leqslant |R(0)|^2, \quad \text{a.e. in } (0, \theta). \tag{4.72}$$

As $S(0) + \tau_h(0) = T^0 - T_r \leqslant |T^0 - T_r|_\infty$, it follows that $R(0) = 0$ and, from (4.72), $|R(t)| = 0$. Recalling (4.71), we see that $\|R(t)\| = 0$, that is $S(t) + \tau_h(t) \leqslant \leqslant C(T^0)$ on $\Omega$ in $H^1(\Omega)$, a.e. in $(0, \theta)$. According to Proposition 5.2.1 in Chapter 1, this gives

$$S(t) + \tau_h(t) \leqslant C(T^0), \quad \text{a.e. in } Q.$$

Similarly, with $R(t) = \min\{S(t) + \tau_h(t) + C(T^0), 0\}$, we obtain

$$S(t) + \tau_h(t) \geqslant -C(T^0), \quad \text{a.e. in } Q.$$

Thus (4.67) is proved, and $S \in L^\infty(Q)$ because

$$\tau_h \in C^2(0, \theta; H^2(\Omega)) \subseteq L^\infty(0, \theta; L^\infty(\Omega)). \qquad \square$$

Let $(\mathbf{u}_1, S_1)$ and $(\mathbf{u}_2, S_2)$ be two solutions of Problem 4.1, corresponding to the initial values $T_1^0$ and $T_2^0$ respectively. With

$$\mathbf{u} = \mathbf{u}_1 - \mathbf{u}_2, \qquad S = S_1 - S_2,$$

we obtain from (4.19) by subtraction

$$(\mathbf{Bu}, \mathbf{v}) + \alpha (S\mathbf{g}, \mathbf{v}) = 0, \quad \forall \mathbf{v} \in \mathbf{H} \text{ a.e. in } (0, \theta). \tag{4.73}$$

From (4.20) we find in the same manner that

$$\sigma \frac{\mathrm{d}}{\mathrm{d}t}(S, T) + (\mathbf{A}\nabla S, \nabla T) + (\mathbf{u}\nabla(S_1 + \tau_h), T) + (\mathbf{u}_2 \nabla S, T) = 0,$$

$$\forall T \in H_0^1(\Omega) \quad \text{a.e. in } (0, \theta). \tag{4.74}$$

Setting $\mathbf{v} = \mathbf{u}$ in (4.73) and $T = S$ in (4.74), from Theorem 4.2 it follows that

$$B_1 |\mathbf{u}| \leqslant \alpha g |S|, \tag{4.75}$$

$$\sigma \frac{\mathrm{d}}{\mathrm{d}t} |S|^2 + 2A_1 \|S\|^2 \leqslant 2K |\mathbf{u}| \|S\| \leqslant A_1 \|S\|^2 + K^2 |\mathbf{u}|^2 / A_1, \tag{4.76}$$

where $K = \min\{C(T_1^0), C(T_2^0)\}$. Recalling the Friedrichs inequality (1.25), from (4.75) and (4.76) we obtain

$$\frac{\mathrm{d}}{\mathrm{d}t} |S|^2 \leqslant C_0 |S|^2, \tag{4.77}$$

where $C_0 = (4K^2 \alpha^2 g^2 L^2 A_1^{-1} B_1^{-2} a_0^{-2} \sigma^{-1} - A_1 \sigma^{-1})$. Therefore

$$|S_1(t) - S_2(t)| \leqslant |T_1^0 - T_2^0| \exp(C_0 t). \tag{4.78}$$

Using (4.75) again, we obtain also

$$|\mathbf{u}_1(t) - \mathbf{u}_2(t)| \leqslant \frac{\alpha g}{B_1} T_1^0 - T_2^0 \exp(C_0 t). \tag{4.79}$$

In particular, with $T_1^0 = T_2^0$ in (4.78) and (4.79), we prove

THEOREM 4.3. *Problem* 4.1 *has a unique solution.*

The relations (4.78) and (4.79) permit us to establish also a stability result:

THEOREM 4.4. *Any perturbation of the initial value (in the sense studied above) decreases exponentially with time if*

$$\min\{C(T_1^0), C(T_2^0)\} \leqslant \frac{a_0}{2} \left(\frac{A_1 B_1}{\alpha g L}\right). \tag{4.80}$$

We now consider the implications of these last results for the steady case.

If the parameters $\mathbf{g}$ and $\tau$ are time-independent, and $S^0$ together with some $\mathbf{u}^0$ is a steady solution of (4.13)–(4.16), then Theorem 4.3 implies that the solution of Problem 4.1 is

$$(\mathbf{u}(t), S(t)) = (u^0, S^0) \quad \text{for a.a. } t \in (0, \theta). \tag{4.81}$$

Now, suppose that two initial values $S_1^0$ and $S_2^0$ ($S_i^0 = T_i^0 - (\tau_h + T_r)$) are eady solutions. Then, following (4.81), we obtain

$$S_1(t) = S_1^0, \quad S_2(t) = S_2^0 \quad \text{for a.a. } t \in (0, \theta);$$

thus (4.78) becomes

$$|T_1^0 - T_2^0| \leqslant |T_1^0 - T_2^0| \exp(C_0 t). \tag{4.82}$$

If $|T_1^0 - T_2^0| \neq 0$ and (4.80) is satisfied, then (4.82) yields a contradiction for sufficiently large $t$. Therefore we can say that (4.80) becomes in the steady case a uniqueness criterion. Unfortunately, no new results emerge because, using the notation of Section 1,

$$|\tau - T_r|_\infty = \beta/2$$

and, recalling the corresponding maximum principle (Theorem 1.3),

$$|S_1^0 + \tau_h|_\infty = |T_i^0 - T_r| \leqslant \beta/2, \quad i = 1, 2.$$

Thus $K = \beta/2$, and therefore (4.80) is exactly (2.1), which is already known.

## 5. Comments and Bibliographical Notes

The existence Theorem 1.1 is given by Poliševski (1985a); it generalizes a rather classical theorem given by Gossez (1966). The characterization of the space $\mathbf{H}^\perp$ can be found in Temam (1977, Chap. 1).

There is no systematic review of the properties of Sobolev spaces; the reader is usually referred to Nečas (1967), Lions and Magenes (1968) and Adams (1975).

The weak maximum principles are formulated in terms of inequalities in the sense of $H^1(\Omega)$. In the definitions and techniques we have followed Kinderlehrer and Stampacchia (1980).

Most of the material in Section 2.1 has appeared in Poliševski (1985 a, c).

The first part of Section 2.2 contains a uniqueness criterion and a study of the case $a < a_0$, which are taken from Poliševski (1985*d*). For the problem of convection between concentric spheres, the reader should consult also Burns and Tien (1979) or Poliševski (1979).

The case $n = 2$ of Lemma 3.3 was proved by Foiaş (1972, Sect. 7). In general, the method used in Section 3 are adopted from Foiaş and Temam (1977). We have to remark here that the main difference between these problems and those raised by the Navier – Stokes equations is that the buoyancy non-linear term of the Darcy – Boussinesq system $(\mathbf{u}\,\nabla \underset{\centerdot}{T})$ is a priori only measurable, while the inertial term of the Navier – Stokes equations $((\mathbf{u}\nabla)\,\mathbf{u})$ belongs to $\mathbf{L}^2(\Omega)$ if $n = 2$, and to $\mathbf{L}^{3/2}(\Omega)$ if $n = 3$; that is why we are concerned so much with regularity properties. For instance, in Section 3 we need all the $H^2(\Omega)$-regularity results from Section 2 and, moreover, the peculiar weak-maximum principle result (Theorem 3.4). The results given in Section 2.3 have appeared in Poliševski (1985e).

In order to prove the existence in the evolution case, we have used the Galerkin procedure; the corresponding weak maximum principle and the uniqueness are then easily obtained. These results were announced and partially proved in Poliševski (1985b); see also Poliševski (1986a). Note that (4.80) is in fact a global stability in the mean criterion, in the sense used by Serrin (1959).

CHAPTER 3

# Natural Convection in Unbounded Domains

If the porous medium is not bounded, no analysis is possible without information on the behaviour of the solution at infinity. From this aspect arise some peculiarities which prevent one from using any result obtained for the bounded case without first carefully checking its validity.

Here we consider problems of natural convection in which the main types of boundary condition at infinity occur. In Section 1, which deals with a horizontal layer, we consider doubly periodic solutions in the horizontal variables. The solutions of the exterior problems considered in Section 2 vanish at infinity. In Section 3, on the domain confined between two horizontal concentric cylinders, we study two dimensional solutions; i.e., solutions independent of the axis of symmetry variable. At high Rayleigh numbers the most important part of the convection takes place in a thin layer around the heating surface. In Section 4 we give a boundary layer approximation similar to that used by Prandtl for unbounded viscous fluids.

## 1. Horizontal Porous Layer

The occurrence of convective motion in a horizontal porous layer bounded by two isothermal planes maintaining a temperature gradient opposite to the direction of gravity has been studied continuously since the first work over forty years ago by Horton and Rogers (1945). The early linear theories (see for instance Lapwood (1948)), determined the criterion for the stability of the conduction state and predicted some features of convective flow at Rayleigh numbers above a certain critical value. From the mathematical point of view, below the critical Rayleigh number the boundary conditions are sufficient to determine the stationary solution uniquely (and this is the static state), while two or more stationary solutions are possible above that value. In fact, the linear theories, which apply only for the infinitesimal amplitude of the phenomenon, predict an infinite degeneracy.

Here we study the Rayleigh number regime near the critical point. We give the finite-amplitude steady solutions, and investigate their stability to superposed infinitesimal disturbances.

### 1.1. THE FUNDAMENTAL EQUATIONS

Let O be the origin of the coordinate system $Oxyz$, so that negative $Oz$ direction is the direction in which the gravitational force acts ($\mathbf{g} = -g\mathbf{e}$, $g > 0$, $\mathbf{e} = (0, 0, 1)$). The domain is bounded at the bottom by the plane $z = 0$ and at the top by the plane $z = L$, $L > 0$. The classical non-stationary equations, similar to (2.34)–(2.36) in Chapter 2, are

$$\operatorname{div} \mathbf{u} = 0, \quad z \in (0, L), \tag{1.1}$$

$$\frac{\mu}{k}\mathbf{u} + \nabla p = \rho_f (1 - \alpha (T - T_r))\,\mathbf{g}, \quad z \in (0, L), \tag{1.2}$$

$$(\rho c)_m\, \partial_t T + \rho_f C_f\, \mathbf{u} \nabla T = \chi \Delta T, \quad z \in (0, L), \tag{1.3}$$

where $(\rho c)_m$ is the specific heat of the porous medium; we use the following notation:

$$\partial_t = \frac{\partial}{\partial t}, \qquad \partial_x = \frac{\partial}{\partial x}, \qquad \partial_y = \frac{\partial}{\partial y}, \qquad \partial_z = \frac{\partial}{\partial z},$$

$$\mathbf{u} = (u, v, w).$$

The corresponding boundary conditions are

$$w = 0, \quad z = 0, z = L, \tag{1.4}$$

$$T = T_1, \quad z = 0, \tag{1.5}$$

$$T = T_2, \quad z = L, \tag{1.6}$$

where $T_1 > T_2$ since the layer is heated from below.

The dimensionless form of the system (1.1)–(1.6) can be obtained by defining

$$t^* = \chi L^{-2} (\rho c)_m^{-1}\, t,$$
$$(x^*, y^*, z^*) = L^{-1} (x, y, z),$$
$$T^* = (T_1 - T_2)^{-1} (T - T_2),$$
$$\mathbf{u}^* = \rho_f C_f\, L \chi^{-1} \mathbf{u},$$
$$h^* = \mu^{-1} \chi^{-1} \rho_f C_f\, k\, [p + \rho_f\, g (1 + \alpha (T_r - T_2))\, z],$$
$$a^* = \mu^{-1} \chi^{-1} \rho_f^2 C_f\, L g \alpha (T_1 - T_2)\, k,$$

where $a^*$ is the dimensionless parameter in the problem, the Rayleigh number.

Omitting the asterisks to simplify the notation, we obtain

$$\operatorname{div} \mathbf{u} = 0, \qquad z \in (0, 1), \tag{1.7}$$

$$\mathbf{u} + \nabla h = aT\mathbf{e}, \qquad z \in (0, 1), \tag{1.8}$$

$$\partial_t T + \mathbf{u} \nabla T = \Delta T, \qquad z \in (0, 1), \tag{1.9}$$

$$w = 0, \qquad z = 0, z = 1, \tag{1.10}$$

$$T = 1\,, \qquad z = 0\,, \tag{1.11}$$

$$T = 0\,, \qquad z = 1\,. \tag{1.12}$$

### 1.2. THE STEADY SOLUTIONS

Any solution of (1.7)–(1.12) with $\partial_t = 0$ is called a steady solution. Obviously the static solution

$$\mathbf{u}_0 = 0\,, \tag{1.13}$$

which corresponds to the pure-conduction pattern

$$T_0 = 1 - z\,, \tag{1.14}$$

$$h_0 = a\left(z - \frac{1}{2}z^2\right) + \text{const.}, \tag{1.15}$$

is a steady solution of (1.7)–(1.12). As in Section 2 of Chapter 2, one can prove that this solution is unique for sufficiently low Rayleigh numbers.

Introducing

$$q = h - h_0\,, \tag{1.16}$$

$$S = T - T_0\,, \tag{1.17}$$

the stationary system (1.7)–(1.12) becomes

$$\operatorname{div}\mathbf{u} = 0, \qquad z \in (0, 1)\,, \tag{1.18}$$

$$\mathbf{u} + \nabla q = aS\mathbf{e}\,, \qquad z \in (0, 1)\,, \tag{1.19}$$

$$\mathbf{u}\nabla S = w + \Delta S\,, \qquad z \in (0, 1)\,, \tag{1.20}$$

$$w = S = 0\,, \qquad z = 0, z = 1\,. \tag{1.21}$$

As we want to study the case of Rayleigh numbers close to the critical value, we assume that the amplitude $\varepsilon$ of $S$, the difference between a steady solution occurring after the critical point and the static-state /pure-conduction solution, is a small parameter. Thus the Rayleigh number and the solutions of (1.18)–(1.21) may be approximated by the formal expansions

$$a = a_0 + \varepsilon a_1 + \varepsilon^2 q_2 + \ldots, \tag{1.22}$$

$$\mathbf{u} = \varepsilon \mathbf{u}_1 + \varepsilon^2 \mathbf{u}_2 + \cdots, \tag{1.23}$$

$$q = \varepsilon q_1 + \varepsilon^2 q_2 + \cdots, \tag{1.24}$$

$$S = \varepsilon S_1 + \varepsilon^2 S_2 + \cdots. \tag{1.25}$$

REMARK 1.1. Since $a$ is an externally given parameter, Equation (1.22) implicitly defines $\varepsilon$.

In terms of the series (1.25), the hypothesis that $\varepsilon$ is the amplitude of $S$ is made explicit in the following way:

$$\int_D S_1^2 \, dx = 1 , \tag{1.26}$$

$$\int_D S_k S_1 \, dx = 0 , \quad \forall\, k \geqslant 2 , \tag{1.27}$$

where the integration is over some characteristic domain $D = Y \times [0,1]$, $Y \subseteq \mathbb{R}^2$. Usually it is assumed that convective motion occurs in a periodic cell pattern, where a cell is defined by a closed boundary with vertical walls between $z = 0$ and $z = 1$, the bottom and top layers. (A periodic cell pattern is a configuration in which any two neighbouring cells are identical.) Moreover, the normal component of the velocity should vanish on the boundary of a cell.

REMARK 1.2. From (1.26) and (1.27), we obtain

$$\| S \|_{L^2(D)} = \varepsilon + O(\varepsilon^3).$$

The coefficients $a_{n-1}$, $\mathbf{u}_n$, $q_n$ and $S_n$ $(n \geqslant 1)$ are found by substituting (1.22)–(1.25) in (1.18)–(1.21) and solving the equations obtained by collecting together terms with the same power of ε.

The first-order problem is

$$\operatorname{div} \mathbf{u}_1 = 0, \qquad z \in (0,1), \tag{1.28}$$

$$\mathbf{u}_1 + \nabla q_1 = a_0 S_1 \mathbf{e} \quad z \in (0,1), \tag{1.29}$$

$$\Delta S_1 + w_1 = 0 , \qquad z \in (0,1), \tag{1.30}$$

$$S_1 = w_1 = 0 , \qquad z = 0, z = 1 . \tag{1.31}$$

Operating with div on (1.29), we find that

$$\Delta q_1 = a_0 \partial_z S_1 . \tag{1.32}$$

Eliminating $w_1$ between (1.29) and (1.30), we get

$$\Delta S_1 - \partial_z q_1 + a_0 S_1 = 0 . \tag{1.33}$$

Operating with Δ on (1.33), and using (1.32), we obtain

$$\Delta^2 S_1 + a_0 \Delta_{xy} S_1 = 0, \quad \Delta_{xy} = \frac{\partial^2}{\partial x^2} + \frac{\partial^2}{\partial y^2} \cdot \tag{1.34}$$

Taking into account (1.30) and (1.31), the boundary conditions for Equation (1.34) are

$$S_1 = \Delta S_1 = 0, \quad z = 0, z = 1 . \tag{1.35}$$

It is obvious that (1.34) and (1.35) represent a characteristic-value problem.

Usually such problems have non-vanishing solutions only if the parameter is one of the terms of an increasing sequence of real positive numbers. The problem now consists of finding the first eigenvalue of (1.33)–(1.34). From physical considerations, we must seek eigenfunctions of the form

$$S_1(x, y, z) = F_1(x, y)\, G(z). \tag{1.36}$$

One finds that (1.34) is separable if $F_1$ satisfies the equation for a vibrating membrane,

$$\Delta_{xy} F_1 + n^2 F_1 = 0 . \tag{1.37}$$

Thus the horizontal section $Y$ of a cell in the convection pattern can have any shape compatible with (1.37); the separation constant $n$ corresponds to the principal mode of vibration of a membrane of that shape, and it will be determined explicitly in the course of the solution.

The problem for $G$ reduces to

$$D^4 G - 2n^2 D^2 G + (n^4 - a_0 n^2) G = 0, \quad z \in (0, 1), \tag{1.38}$$

$$G = D^2 G = 0, \quad z = 0, z = 1, \tag{1.39}$$

where $D$ denotes $\mathrm{d}/\mathrm{d}z$.

Considering the characteristic equation of (1.38), and setting

$$\alpha^2 = n^2 - n\sqrt{a_0}\,, \qquad \beta^2 = n^2 + n\sqrt{a_0}\,, \tag{1.40}$$

we can write $G$ in the form

$$G(z) = C_1 \mathrm{e}^{\alpha z} + C_2 \mathrm{e}^{-\alpha z} + C_3 \mathrm{e}^{\beta z} + C_4 \mathrm{e}^{-\beta z} .$$

In order to satisfy the boundary conditions (1.39), we have to impose the additional conditions

$$\mathrm{e}^{\alpha} C_1 + \mathrm{e}^{-\alpha} C_2 + \mathrm{e}^{\beta} C_3 + \mathrm{e}^{-\beta} C_4 = 0,$$

$$C_1 + C_2 + C_3 + C_4 = 0 ,$$

$$-\mathrm{e}^{\alpha} C_1 - \mathrm{e}^{-\alpha} C_2 + \mathrm{e}^{\beta} C_3 + \mathrm{e}^{-\beta} C_4 = 0 ,$$

$$-C_1 - C_2 + C_3 + C_4 = 0 .$$

As $\beta$ is real and non-zero, we have immediately

$$C_3 = C_4 = 0 .$$

Since we are looking for non-trivial values of $G$ and $C_1 = -C_2$, it is necessary that $\alpha \neq 0$ and $\mathrm{e}^{\alpha} - \mathrm{e}^{-\alpha} = 0$; that is,

$$\alpha = \mathrm{i} k \pi, \quad k \in \mathbb{Z} \setminus \{0\} .$$

Via (1.40) we obtain $a_0$ as

$$a_0 = \left( n + \frac{k^2 \pi^2}{n} \right)^2 , \quad k \in \mathbb{Z} \setminus \{0\} .$$

Obviously, the smallest value of $a_0$ is obtained for

$$k^2 = 1 , \qquad n = \pi . \tag{1.41}$$

From now on all the computations are straightforward, and lead to:

$$a_0 = 4\pi^2 , \tag{1.42}$$

$$S^1 = F_1(x, y) \sin \pi z , \tag{1.43}$$

$$q_1 = - 2\pi F_1(x, y) \cos \pi z , \tag{1.44}$$

$$u_1 = 2\pi \, \partial_x F_1(x, y) \cos \pi z , \tag{1.45}$$

$$v_1 = 2\pi \, \partial_y F_1(x, y) \cos \pi z , \tag{1.46}$$

$$w_1 = 2\pi^2 F_1(x, y) \sin \pi z , \tag{1.47}$$

where $F_1$ satisfies the conditions

$$\Delta_{xy} F_1 + \pi^2 F_1 = 0 , \tag{1.48}$$

$$\int_Y F_1^2 (x, y) \, \mathrm{d}x \, \mathrm{d}y = 2 , \tag{1.49}$$

$$\frac{\partial F_1}{\partial \mathbf{n}} = 0 \text{ on the boundary of } Y, \tag{1.50}$$

$\mathbf{n}$ being the normal on $\partial Y$.

REMARK 1.3. The condition (1.49) is equivalent to (1.26), and the condition (1.50) follows from (1.45) and (1.46) and from the definition of the cell.

Now we can obtain step by step the other unknown coefficients. If $a_{k-1}$, $\mathbf{u}_k$, $q_k$ and $S_k$ are determined for any $k < n$ $(n \geqslant 2)$, then we find the $n$th-order unknowns from the problem

$$\operatorname{div} \mathbf{u}_n = 0, \quad z \in (0, 1) , \tag{1.51}$$

$$\mathbf{u}_n + \nabla q_n = 4\pi^2 S_n \mathbf{e} + \sum_{k=1}^{n-1} a_k S_{n-k} , \quad z \in (0, 1) , \tag{1.52}$$

$$\Delta S_n + w_n = \sum_{k=1}^{n-1} \mathbf{u}_k \nabla S_{n-k} , \quad z \in (0, 1) , \tag{1.53}$$

$$w_n = S_n = 0, \quad z = 0, z = 1. \tag{1.54}$$

By a process similar to (1.32)–(1.35), we obtain

$$\Delta^2 S_n + 4\pi^2 \Delta_{xy} S_n = \sum_{k=1}^{n-1} \left[ \Delta (\mathbf{u}_k \nabla S_{n-k}) - a_k \Delta_{xy} S_{n-k} \right] , \tag{1.55}$$

$$S_n = \Delta S_n - \sum_{k=1}^{n-1} \mathbf{u}_k \nabla S_{n-k} = 0 \quad z = 0, z = 1 . \tag{1.56}$$

Let us consider the second-order problem

$$\Delta^2 S_2 + 4\pi^2 \Delta_{xy} S_2 = \Delta\,(\mathbf{u}_1\ \nabla S_1) - a_1 \Delta_{xy} S_1\,, \tag{1.57}$$

$$S_2 = \Delta\, S_2 - (\mathbf{u}_1\ \nabla S_1) = 0, \quad z = 0, z = 1\,. \tag{1.58}$$

As $(\mathbf{u}_1 \nabla S_1) = \pi \sin 2\pi z\,((\partial_x F_1)^2 + (\partial_y F_1)^2 + \pi^2 F_1^2)$, and hence $\Delta S_2 = 0$ for $z = 0$ and $z = 1$, by multiplying (1.57) with $S_1$ and integrating over $D$ we obtain

$$0 = \int_D (\Delta^2 S_2 + 4\pi^2\,\Delta_{xy} S_2)\, S_1 + \left[a_1 \Delta_{xy}\, S_1 - \Delta\,(\mathbf{u}_1 \nabla S_1)\right] S_1$$

$$= \int_D S_2\,(\Delta^2 S_1 + 4\pi^2 \Delta_{xy} S_1) - a_1 \pi^2 S_1^2 - (\mathbf{u}_1 \nabla S_1)\,\Delta S_1 =$$

$$= -\ a_1 \pi^2 + 2\pi^2 \int_D \mathbf{u}_1\ S_1 \Delta S_1 = -\ a_1 \pi^2 + \pi^2 \int_D \operatorname{div}\,(\mathbf{u}_1 S_1^2) =$$

$$= -\ a_1 \pi^2\,, \tag{1.59}$$

where we have used also (1.58), (1.26) and (1.27), and the $Y$-periodicity of $\mathbf{u}_1$ and $S_1$. Thus we have proved that

$$a_1 = 0\,, \tag{1.60}$$

the problem for $S_2$ reducing to

$$\Delta^2 S_2 + 4\pi^2 \Delta_{xy} S_2 = \pi \sin 2\pi z\,(\Delta_{xy}\, G_1 - 4\pi^2 G_1)\,, \tag{1.61}$$

$$S_2 = \Delta\, S_2 = 0, \quad z = 0, z = 1\,, \tag{1.62}$$

where $G_1(x, y) = (\partial_x F_1)^2 + (\partial_y F_1)^2 + \pi^2 F_1^2$. If we look now for a particular solution of (1.61) and (1.62) of the form $\pi \sin 2\pi z F_2(x, y)$, we immediately obtain the equation

$$(\Delta_{xy}^2 - 4\pi^2 \Delta_{xy} + 16\pi^4) F_2(x, y) = \Delta_{xy} G_1 - 4\pi^2 G_1\,, \tag{1.63}$$

from which a $Y$-periodic solution $F_2$ is determined uniquely. Therefore the general solution of (1.61) and (1.62) is

$$S_2 = \pi \sin 2\pi z F_2(x, y) + C_1 S_1\,, \tag{1.64}$$

where the constant $C_1$ is given by the condition (1.27):

$$\int_D S_2 S_1\, \mathrm{d}x = 0\,.$$

From now on nothing new will appear in the procedure for determining the coefficients. However, we think that is interesting to give the formula for $a_{n-1}$ ($n \geqslant 3$), which is obtained by multiplying (1.55) by $S_1$ and integrating over $D$, as in (1.59):

$$a_{n-1} = 2 \sum_{k=1}^{n-2} \int_D \mathbf{u}_k S_1 \nabla S_{n-k} \, . \tag{1.65}$$

The increase in the mean heat transfer, resulting from convection, cannot be described by the linear theory because any product of disturbance components is omitted. The method used here permits the Nusselt number to be computed as

$$Nu = \int_Y \frac{\partial T}{\partial z}\bigg|_{z=1} \mathrm{d}x \, \mathrm{d}y \bigg/ \int_Y \frac{\partial T_0}{\partial z}\bigg|_{z=1} \mathrm{d}x \, \mathrm{d}y \, . \tag{1.66}$$

Then, using (1.17) and (1.25), we have

$$Nu = 1 - \varepsilon \int_Y \frac{\partial S_1}{\partial z}\bigg|_{z=1} \mathrm{d}x \, \mathrm{d}y - \varepsilon^2 \int_Y \frac{\partial S_2}{\partial z}\bigg|_{z=1} \mathrm{d}x \, \mathrm{d}y + O(\varepsilon^3) \, .$$

By straight integration of (1.48) over $Y$, we obtain

$$\int_Y F_1 (x, y) \, \mathrm{d}x \, \mathrm{d}y = 0 \, . \tag{1.67}$$

and hence

$$Nu = 1 - 2\pi^2 \varepsilon^2 \int_Y F_2 (x, y) \, \mathrm{d}x \, \mathrm{d}y + O(\varepsilon^3). \tag{1.68}$$

Multiplying (1.49) by $F_1$ and integrating over $Y$, we obtain via (1.49)

$$\int_Y (\partial_x F_1)^2 + (\partial_y F_1)^2 \, \mathrm{d}x \, \mathrm{d}y = \int_Y \pi^2 F_1^2 \, \mathrm{d}x \, \mathrm{d}y = 2\pi^2 \, ,$$

and consequently

$$\int_Y G_1 (x, y) \, \mathrm{d}x \, \mathrm{d}y = 4\pi^2 \, \cdot \tag{1.69}$$

Finally, integrating (1.63) over $Y$ and using the $Y$-periodicity of $F_2$ and $G_1$, we get

$$16\pi^2 \int_Y F_2 (x, y) \, \mathrm{d}x \, \mathrm{d}y = -4\pi^2 \int_Y G_1 (x, y) \, \mathrm{d}x \, \mathrm{d}y \, ,$$

and from (1.69) it follows that

$$\int_Y F_2 (x, y) \, \mathrm{d}x \, \mathrm{d}y = -1 \, . \tag{1.70}$$

Introducing (1.70) in (1.68), we obtain

$$Nu = 1 + 2\pi^2\varepsilon^2 + O(\varepsilon^3). \tag{1.71}$$

For the third-order approximation of (1.22), we have

$$a = 4\pi^2 + a_2\varepsilon^2 + O(\varepsilon^3), \tag{1.72}$$

where $a_2$ can be computed by using (1.65). Using (1.43), (1.45) and (1.64), we have

$$a_2 = -\pi^2 \int_Y G_1(x, y)\, F_2(x, y)\, dx\, dy. \tag{1.73}$$

Then, for a given Rayleigh number $a$, not far from the critical point $4\pi^2$, the heat transfer is described for the third-order approximation by

$$Nu = 1 + 2\pi^2 (a - 4\pi^2)/a_2. \tag{1.74}$$

From the above analyses and results, it is apparent that the first question we must ask is how close the eigenfunctions (1.36) and (1.37) are to the first eigenfunctions of (1.34) and (1.35). The only answer we can give is that many experimentalists have confirmed $a_0 = 4\pi^2$ to be the critical Rayleigh number. Next, even accepting (1.36) and (1.37), the $Y$-periodic pattern of (1.48)–(1.50) is not well-specified. Since there are no preferred points or directions in the horizontal sections, it seems reasonable to assume that the entire horizontal plane is covered by regular polygons – equilateral triangles, squares or regular hexagons.

For the square pattern of convection, we have

$$F_1(x, y) = \cos(\pi x/\sqrt{2}) \cos(\pi y/\sqrt{2}),$$

$$Y = [-\sqrt{2}, \sqrt{2}] \times [-\sqrt{2}, \sqrt{2}],$$

which are determined uniquely by (1.48)–(1.50). Then

$$G_1(x, y) = \pi^2(2 + \cos(\pi x\sqrt{2}) + \cos(\pi y\sqrt{2}))/4,$$

and the only $Y$-periodic solution of (1.63) is

$$F_2(x, y) = -(7 + 3\cos(\pi x\sqrt{2} + 3)\cos(\pi y\sqrt{2}))/56.$$

Finally we obtain

$$a_2 = 17\pi^4/28,$$

and for the third-order approximation

$$\varepsilon \simeq \frac{2\sqrt{119}}{17\pi^2}\sqrt{a - 4\pi^2},$$

$$Nu \simeq 1 + \frac{56}{17\pi^2}(a - 4\pi^2).$$

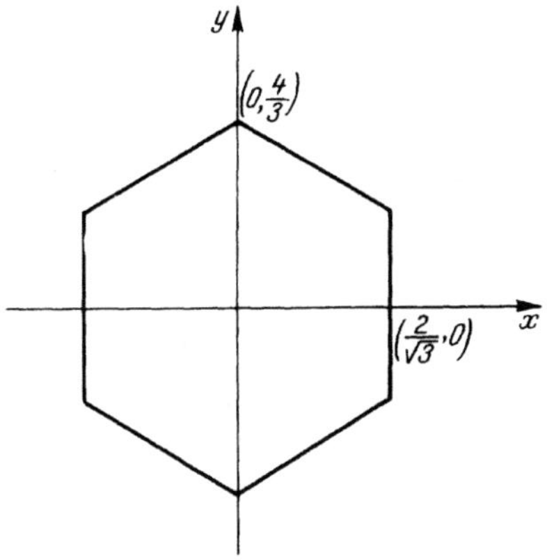


Fig. 9. Hexagon pattern.

The regular hexagonal pattern is obtained for an edge length of 4/3 (see Fig. 9), with

$$F_1(x, y) = 2 \cos \frac{\pi\sqrt{3}}{2} x \cos \frac{\pi}{2} y + \cos \pi y .$$

Another interesting case is shown in Fig. 10, that of the rolls pattern (two-dimensional with $\partial_y = v = 0$). It needs separate treatment because Equation (1.49) does not make sense for $Y = [-1, 1] \times \mathbb{R}$; it have to be replaced by

$$\int_{-1}^{1} F_1^2(x)\,\mathrm{d}x = 2 . \tag{1.49'}$$

Thus $F_1(x) = \sqrt{2} \cos \pi x$, (see Fig. 10), $G_1(x) = 2\pi^2$, $F_2(x) = -\dfrac{1}{2}$ and

$$a_2 = -\pi^2 \int_{-1}^{1} G_1(x)\, F_2(x)\,\mathrm{d}x = 2\pi^4 .$$

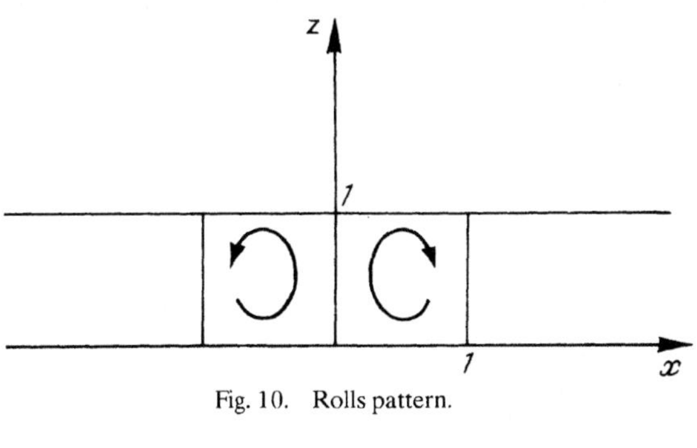


Fig. 10. Rolls pattern.

1.3. THE STABILITY ANALYSIS

If we superpose infinitesimal disturbances on the steady solutions; that is, if we take in (1.7)−(1.12)

$$\mathbf{u}(x, y, z, t) = \tilde{\mathbf{u}}(x, y, z)\exp(\sigma t) + \mathbf{u}(x, y, z),$$
$$T(x, y, z, t) = \tilde{S}(x, y, z)\exp(\sigma t) + (1 - z) + S(x, y, z),$$
$$h(x, y, z, t) = \tilde{q}(x, y, z)\exp(\sigma t) + a\left(z - \frac{1}{2}z^2\right) + q(x, y, z),$$

where $\mathbf{u}$, $S$ and $q$ satisfy (1.18)−(1.21), and if we neglect the disturbance products, then we obtain the stability equations of the solution $(\mathbf{u}, S, q)$ as

$$\operatorname{div}\tilde{\mathbf{u}} = 0, \quad z \in (0, 1), \tag{1.75}$$
$$\tilde{\mathbf{u}} + \nabla\tilde{q} = a\tilde{S}\mathbf{e}, \quad z \in (0, 1), \tag{1.76}$$
$$\sigma\tilde{S} + \tilde{\mathbf{u}}\nabla S + \mathbf{u}\nabla\tilde{S} = \Delta\tilde{S} + \tilde{w}, \quad z \in (0, 1), \tag{1.77}$$
$$\tilde{w} = \tilde{S} = 0, \quad z = 0, z = 1, \tag{1.78}$$

where the components of $\tilde{\mathbf{u}}$ are $(\tilde{u}, \tilde{v}, \tilde{w})$.

The system (1.75)−(1.78) is obviously a characteristic-value problem for the growth rate $\sigma$. In accordance with our previous assumption that $\varepsilon$ is the small parameter in the problem, we approximate the growth rate $\sigma$ and the infinitesimal disturbances $\tilde{\mathbf{u}}$, $\tilde{S}$ and $\tilde{q}$ by the formal expansions

$$\sigma = \sigma_0 + \varepsilon\sigma_1 + \varepsilon^2\sigma_2 + \cdots, \tag{1.79}$$
$$\tilde{\mathbf{u}} = \tilde{\mathbf{u}}_1 + \varepsilon\tilde{\mathbf{u}}_2 + \cdots, \tag{1.80}$$
$$\tilde{q} = \tilde{q}_1 + \varepsilon\tilde{q}_2 + \cdots, \tag{1.81}$$
$$\tilde{S} = \tilde{S}_1 + \varepsilon\tilde{S}_2 + \cdots. \tag{1.82}$$

These series, together with the series (1.22)−(1.25), when substituted into the system (1.75)−(1.78) yield, after application of the usual perturbation method, a set of equations from which for any $n \geqslant 1$ the eigenvalues $\sigma_{n-1}$ and the eigenfunctions $\tilde{\mathbf{u}}_n$, $\tilde{q}_n$ and $\tilde{S}_n$ can be obtained in a straightforward manner. The stability analysis is specified by:

DEFINITION 1.1. The steady solution $(\mathbf{u}, q, S)$ is stable if all possible values of $\sigma$ are less than zero.

The system obtained by collecting together the $\varepsilon^0$-terms is

$$\operatorname{div}\tilde{\mathbf{u}}_1 = 0, \quad z \in (0, 1), \tag{1.83}$$
$$\tilde{\mathbf{u}}_1 + \nabla\tilde{q}_1 = 4\pi^2\tilde{S}_1\mathbf{e}, \quad z \in (0, 1), \tag{1.84}$$
$$\sigma_0\tilde{S}_1 = \Delta\tilde{S}_1 + \tilde{w}_1, \quad z \in (0, 1), \tag{1.85}$$
$$\tilde{w}_1 = \tilde{S}_1 = 0, \quad z = 0, \ z = 1. \tag{1.86}$$

Eliminating as usual $\tilde{\mathbf{u}}_1$ and $\tilde{q}_1$, we find that

$$\Delta^2\tilde{S}_1 + 4\pi^2\Delta_{xy}\tilde{S}_1 = \sigma_0\Delta\tilde{S}_1, \qquad z \in (0,1), \tag{1.87}$$

$$\tilde{S}_1 = \Delta\tilde{S}_1 = 0, \qquad z = 0, z = 1. \tag{1.88}$$

This is a characteristic-value problem similar to that given by (1.34) and (1.35), and we look for eigenfunctions of the form

$$\tilde{S}_1 = \tilde{F}_1(x, y)\,\tilde{G}(z). \tag{1.89}$$

Again, one finds that (1.87) is separable if $\tilde{F}_1$ satisfies the equation for a vibrating membrane:

$$\Delta_{xy}\tilde{F}_1 + m^2\tilde{F}_1 = 0, \tag{1.90}$$

where $m$ is an arbitrary wavenumber. The problem for $\tilde{G}$ reduces to

$$D^4\tilde{G} - (2m^2 + \sigma_0)D^2\tilde{G} + (m^4 - 4\pi^2m^2 + \sigma_0 m^2)\tilde{G} = 0, \tag{1.91}$$

$$\tilde{G} = D^2\tilde{G} = 0, \quad z = 0, z = 1. \tag{1.92}$$

The roots of the characteristic equation of (1.91), $\pm\,\alpha$ and $\pm\,\beta$, are given by

$$\alpha^2 = (2m^2 + \sigma_0 - \sqrt{\sigma_0^2 + 16\pi^2m^2})/2,$$

$$\beta^2 = (2m^2 + \sigma_0 + \sqrt{\sigma_0^2 + 16\pi^2m^2})/2.$$

As $\beta$ is real and non-zero, it follows as usual that a non-trivial $\tilde{G}$ exists only if

$$\alpha = ik\pi, \quad k \in \mathbb{Z}\setminus\{0\}.$$

It follows immediately that

$$\sigma_0(m^2 + k^2\pi^2) = -(m^2 - 2\pi m + k^2\pi^2)(m^2 + 2\pi m + k^2\pi^2),$$

from which $\sigma_0 < 0$, except when $k^2 = 1$ and $m = \pi$, when $\sigma_0 = 0$.

This means that all the first-order disturbances lead to stability, except those which have the same wave number as in the steady solutions; these disturbances are in the irresolute case $\sigma_0 = 0$. Therefore in this case we have to compute higher-order terms.

First, we note that if $\sigma_0 = 0$ then (1.83)−(1.86) is the same as (1.28)−(1.31), and hence we have the eigenfunctions

$$(\tilde{\mathbf{u}}_1, \tilde{q}_1, S_1) = (\mathbf{u}_1, q_1, S_1), \tag{1.93}$$

where $\mathbf{u}_1$, $q_1$ and $S_1$ are given by (1.43)−(1.47).

The corresponding second-order system is

$$\operatorname{div}\tilde{\mathbf{u}}_2 = 0, \qquad z \in (0,1), \tag{1.94}$$

$$\tilde{\mathbf{u}}_2 + \nabla\tilde{q}_2 = 4\pi^2\tilde{S}_2\mathbf{e}, \qquad z \in (0,1), \tag{1.95}$$

$$\sigma_1 S_1 + 2\mathbf{u}_1\nabla S_1 = \Delta\tilde{S}_2 + \tilde{w}_2, \qquad z \in (0,1), \tag{1.96}$$

$$\tilde{w}_2 = \tilde{S}_2 = 0, \qquad z = 0, z = 1. \tag{1.97}$$

As in (1.59), one can prove that $\sigma_1 = 0$, and consequently

$$\tilde{S}_2 = 2S_2, \quad \tilde{\mathbf{u}}_2 = 2\mathbf{u}_2, \tilde{q}_2 = 2q_2, \tag{1.98}$$

where $\mathbf{u}_2$ and $q_2$ correspond to $S_2$ as given by (1.64).

As this result is still indecisive we proceed to consider the third-order system

$$\operatorname{div} \tilde{\mathbf{u}}_3 = 0, \tag{1.99}$$

$$\tilde{\mathbf{u}}_3 + \nabla \tilde{q}_3 = 4\pi^2 \tilde{S}_3 \mathbf{e} + a_2 S_1 \mathbf{e}, \tag{1.100}$$

$$\sigma_2 S_1 + 3\mathbf{u}_2 \nabla S_1 + 3\mathbf{u}_1 \nabla S_2 = \Delta \tilde{S}_3 + \tilde{w}_3, \tag{1.101}$$

$$\tilde{w}_3 = \tilde{S}_3 = 0, \quad z = 0, z = 1. \tag{1.102}$$

Eliminating $\tilde{\mathbf{u}}_3$ and $\tilde{q}_3$, we obtain

$$\Delta^2 \tilde{S}_3 + 4\pi^2 \Delta_{xy} \tilde{S}_3 = \pi^2 (a_2 - 2\sigma_2) S_1 + 3\Delta (\mathbf{u}_2 \nabla S_1 + \mathbf{u}_1 \nabla S_2), \tag{1.103}$$

$$\tilde{S}_3 = \Delta \tilde{S}_3 - 3(\mathbf{u}_2 \nabla S_1 + \mathbf{u}_1 \nabla S_2) = 0, \quad z = 0, z = 1. \tag{1.104}$$

Multiplying (1.103) by $S_1$ and integrating over $D$ we have

$$\pi^2 (a_2 - 2\sigma_2) = \int_D S_1 \Delta \left[\Delta \tilde{S}_3 - 3(\mathbf{u}_2 \nabla S_1 + \mathbf{u}_1 \nabla S_2)\right] +$$

$$+ 4\pi^2 \int_D S_1 \Delta_{xy} \tilde{S}_3 = -2\pi^2 \int_D S_1 \left[\Delta \tilde{S}_3 - 3(\mathbf{u}_2 \nabla S_1 + \mathbf{u}_1 \nabla S_2)\right] -$$

$$- 4\pi^4 \int_D S_1 \tilde{S}_3 = 4\pi^4 \int_D S_1 \tilde{S}_3 + 6\pi^2 \int_D \mathbf{u}_1 S_1 \nabla S_2 -$$

$$- 4\pi^4 \int_D S_1 \tilde{S}_3 = 3\pi^2 a_2, \tag{1.105}$$

where we have used (1.65) for $n = 3$, and the following properties of $S_1$:

$$\Delta S_1 = -2\pi^2 S_1, \tag{1.106}$$

$$\Delta_{xy} S_1 = -\pi^2 S_1. \tag{1.107}$$

From (1.105) it follows that $\sigma_2 = -a_2$; that is, for the marginal stability

$$\sigma = -a_2 \varepsilon^2 + O(\varepsilon^3). \tag{1.108}$$

As any cellular motion which occurs will increase the heat transfer (that is, via (1.71), $a_2 > 0$) we can conclude now that any infinitesimal disturbances of the steady solutions lead to stability.

In view of the 'similarity' between the filtration motion and the motion of the viscous fluid, we might well ask why the first is stable, while the second is clearly unstable. It seems that the principal stabilizing effect in porous media is that inertial effects are far less important; we have ignored them because the corresponding Reynolds number is very small.

This suggests also that a more appropriate stability criterion must be found, for instance that of 'maximum heat transport'.

In the third-order approximation used in this section the maximum heat-transport criterion reduces to the following variational problem: to find $Y$, tessellating $\mathbb{R}^2$, bounded in at least one direction, and the corresponding $F_1$, defined by (1.48)–(1.50), such that $a_2$ given by (1.73) is minimum. It is obvious then, that the square pattern is more stable then the rolls pattern. Straightforward comparisons can be made with the hexagon pattern.

## 2. Exterior Problems

We consider two simple but typical examples of natural convection in unbounded, fluid-saturated porous media in which the induced flow is slow and hence conduction dominates over convection. The convection effects are discussed by using the second-order corrections of the asymptotic solutions of the temperature and velocity field.

The first problem is that of convection about a slightly heated sphere. We start with the system (2.40) and (2.41) in Chapter 2, and obtain the fundamental set of equations

$$\operatorname{div} \mathbf{u} = 0 , \qquad r \geqslant 1 , \tag{2.1}$$

$$\mathbf{u} + \nabla h = aT\mathbf{e}_3 , \quad r \geqslant 1 , \tag{2.2}$$

$$\mathbf{u}\nabla T = \Delta T, \qquad r \geqslant 1 , \tag{2.3}$$

with the boundary conditions

$$\mathbf{u}\cdot\mathbf{v} = 0 , \quad T = 1 , \quad r = 1 , \tag{2.4}$$

$$|\mathbf{u}| \to 0 , \quad T \to 0 \quad \text{as } r \to \infty . \tag{2.5}$$

The analysis can be carried out by straightforward expansions in terms of the small parameter $a$, the Rayleigh number. The successive problems obtained by the perturbation method have unique solutions; in the present case they are independent of the angular direction around the vertical axis. In the spherical polar coordinate system $(r, \theta, \varphi)$, and introducing the stream function $\psi$ as in Chapter 2 ((2.63)), we obtain

$$\frac{1}{\sin\theta}\frac{\partial^2\psi}{\partial r^2} + \frac{1}{r^2}\frac{\partial}{\partial\theta}\left(\frac{1}{\sin\theta}\frac{\partial\psi}{\partial\theta}\right) = a\left(\cos\theta\,\frac{\partial T}{\partial\theta} + r\sin\theta\,\frac{\partial T}{\partial r}\right), \; r \geqslant 1, \tag{2.6}$$

$$\sin\theta\,\frac{\partial}{\partial r}\left(r^2\frac{\partial T}{\partial r}\right) + \frac{\partial}{\partial\theta}\left(\sin\theta\,\frac{\partial T}{\partial\theta}\right) = \frac{\partial T}{\partial r}\frac{\partial\psi}{\partial\theta} - \frac{\partial T}{\partial\theta}\frac{\partial\psi}{\partial r}, \quad r \geqslant 1 , \tag{2.7}$$

$$T = 1 , \quad \psi = 0, \quad r = 1 , \tag{2.8}$$

$$T \to 0\,, \quad r^{-2}\psi \to 0 \quad \text{as } r \to \infty\,. \tag{2.9}$$

Taking $(\psi, T)$ in the form

$$(\psi, T) = \sum_{m \geqslant 0} (\psi_m, T_m)\, a^m\,, \tag{2.10}$$

we find

$$\psi_0 = 0, \qquad T_0 = r^{-1} \tag{2.11}$$

$$\psi_1 = (r - r^{-1}) \sin^2 \theta/2, \qquad T_1 = (2r^{-1} - 3\,r^{-2} + r^{-3}) \cos\theta/4\,, \tag{2.12}$$

$$\psi_2 = (4r - 9 + 6r^{-1} - r^{-2}) \sin^2\theta \cos\theta/24,$$

$$T_2 = \left(-\frac{13}{180} r^{-1} + \frac{11}{240} r^{-3} \ln r + \frac{31}{224} r^{-3} - \frac{13}{144} r^{-4} + \frac{27}{1120} r^{-5}\right) +$$

$$+ \left(\frac{5}{48} r^{-1} - \frac{3}{8} r^{-2} + \frac{11}{80} r^{-3} \ln r + \frac{223}{672} r^{-3} - \frac{1}{12} r^{-4} + \frac{5}{224} r^{-5}\right) \cos 2\theta\,. \tag{2.13}$$

The total quantity of heat necessary to maintain the steady state is characterized by the Nusselt number

$$Nu = \int_{S_1} \frac{\partial T}{\partial r}\bigg|_{r=1} d\sigma \bigg/ \int_{S_1} \frac{\partial T_0}{\partial r}\bigg|_{r=1} d\sigma\,, \tag{2.14}$$

where $S_1$ is the surface of the unit sphere. Explicitly, the Nusselt number is the ratio of the total heat transfer to the heat transfer by conduction alone. It is easily shown that

$$Nu = 1 + \frac{43}{720} a^2 + O(a^3) \tag{2.15}$$

The effect of convection on heat flux is seen to be of order of $a^2$, and is very small. This explains why porous media are so widely used in the thermal insulation of heated bodies.

Next we shall investigate the streamline pattern, considering its second order approximation

$$\psi \simeq a\psi_1 + a^2\psi_2\,. \tag{2.16}$$

As in Section 2 of Chapter 2, we use (2.16) to find the streamlines on which $\psi = 0$, and we find

$$r = 1, \qquad \theta = 0,\ \pi \tag{2.17}$$

and the $(r, \theta)$ points which satisfy

$$\cos\theta = -\frac{12}{a}\left(\frac{r^2 + r}{4r^2 - 5r + 1}\right). \tag{2.18}$$

Therefore we have two kinds of pattern:

(a) If $a \in (0, 3)$, then the pattern is unicellular. The right-hand side of Fig. 11 shows the streamlines, while the left-hand side shows the isothermal lines for $a = 1$. We remark that the warmer region from above is moved upwards by convection.

(b) If $a > 3$, then in the relatively stagnant and cold region below the sphere there appears a second cell (Fig. 12 shows the pattern for $a = 6$). This is as expected, because in the first case the parts of the streamlines below the sphere move off the symmetry axis.

The second problem is that of free convection around a concentrated heat source suddenly embedded in an infinite, fluid-saturated porous medium. This non-stationary problem presents an important reduction in the number of variables, from four ($x$, $y$, $z$ and $t$) to only two. In spite of its relative simplicity, the analysis of this problem is relevant to many practical situations. The transient flow and temperature patterns are derived in the limit of a small parameter – the Rayleigh number, defined by the rate of heat generation at the source.

The phenomenon is governed by the fundamental system of equations (1.1)–(1.3), valid throughout the medium except at the origin of the coordinate system, where the source is situated. As it seems natural to suppose that the problem is symmetrical in the angular direction around the vertical axis passing through the source, we use spherical polar coordinates. This simplifies the problem immediately since neither $\varphi$ nor the velocity component in the $\varphi$-direction appear in the analysis. Thus (1.1) becomes

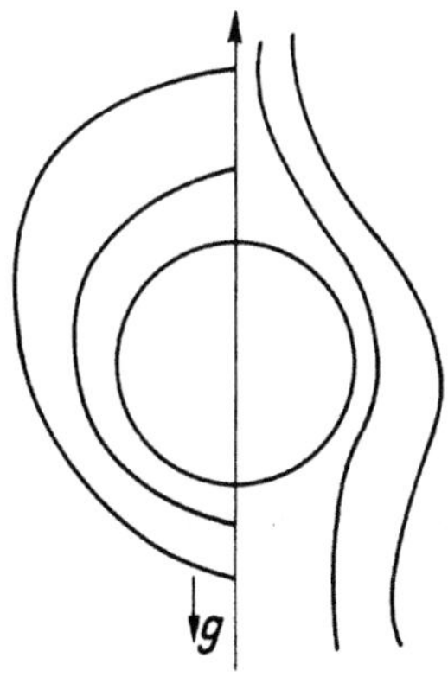


Fig. 11. Streamlines and isothermal lines for $a = 1$.

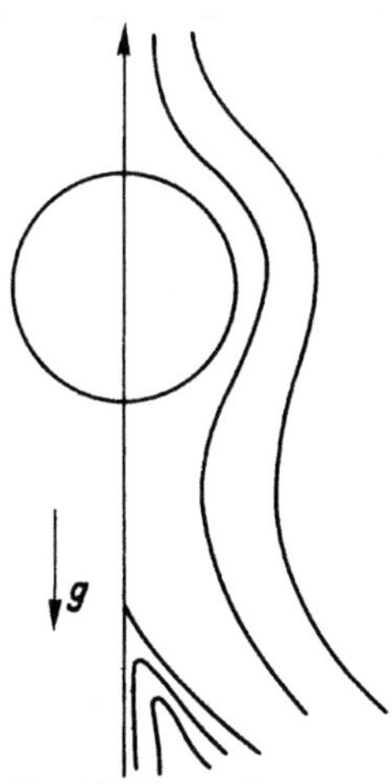


Fig. 12. Streamline pattern for $a = 6$.

$$\frac{\partial}{\partial r}(r^2 \sin\theta\, u_r) + \frac{\partial}{\partial\theta}(r \sin\theta\, u_\theta) = 0 \tag{2.19}$$

and, as usual, a stream function $\psi$ is defined such that (2.19) is satisfied identically:

$$u_r = \frac{1}{r^2 \sin\theta} \frac{\partial\psi}{\partial\theta}, \tag{2.20}$$

$$u_\theta = \frac{-1}{r\sin\theta} \frac{\partial\psi}{\partial r}. \tag{2.21}$$

Introducing also

$$S = T - T_r, \tag{2.22}$$

and eliminating the pressure by operating with curl upon (1.2), from (1.2) and (1.3) we get the following system for $\psi$ and $S$:

$$\frac{1}{r^2}\frac{\partial}{\partial\theta}\left(\frac{1}{\sin\theta}\frac{\partial\psi}{\partial\theta}\right) + \frac{1}{\sin\theta}\frac{\partial^2\psi}{\partial r^2} =$$

$$\mu^{-1} k \rho_f g\alpha\left(\cos\theta \frac{\partial S}{\partial\theta} + r\sin\theta\frac{\partial S}{\partial r}\right), \tag{2.23}$$

$$(\rho c)_m \frac{\partial S}{\partial t} + \frac{\rho_f C_f}{r^2 \sin\theta}\frac{\partial(S,\psi)}{\partial(r,\theta)} =$$

$$\frac{\chi}{r^2\sin\theta}\left[\frac{\partial}{\partial r}\left(r^2\sin\theta\frac{\partial S}{\partial r}\right) + \frac{\partial}{\partial\theta}\left(\sin\theta\frac{\partial S}{\partial\theta}\right)\right], \tag{2.24}$$

in which

$$\frac{\partial(S,\psi)}{\partial(r,\theta)} = \frac{\partial S}{\partial r}\frac{\partial\psi}{\partial\theta} - \frac{\partial S}{\partial\theta}\frac{\partial\psi}{\partial r}$$

is the usual notation for the Jacobian. The system given by (2.23) and (2.24) must be solved subject to the corresponding initial conditions

$$\mathbf{u} = 0, \quad T = T_r, \quad t = 0$$

or

$$\nabla\psi = 0, \quad S = 0, \quad t = 0. \tag{2.25}$$

We have also to describe the asymptotic behaviour far from the source:

$$\mathbf{u} \to 0, \quad T \to T_r, \quad \text{as } |\mathbf{x}| \to \infty$$

or

$$\frac{1}{r}|\nabla\psi| \to 0, \quad S \to 0, \quad \text{as } r \to \infty. \tag{2.26}$$

The fact that we have a concentrated heat source at the origin is described by

$$\int_{S_r}\left(-\chi\frac{\partial T}{\partial r}\right)d\sigma \to q, \quad \text{as } r \to 0\,, \tag{2.27}$$

which represents the heat balance over a spherical surface of radius zero, centred at the origin; $q$ is the strength of the source.

The symmetric assumption implies that the vertical axis passing through the source is a streamline of the flow induced by convection. Without loss of generality, we suppose that its equation is $\psi = 0$. Thus the asymptotic behaviour at the origin is described by

$$\psi \to 0 \quad \text{as } r \to 0\,. \tag{2.28}$$

Also, on the symmetry axis there is only radial movement ($u_\theta = 0$), and naturally, we find there extreme temperatures. Therefore we have the following additional boundary conditions:

$$\frac{1}{\sin\theta}\frac{\partial\psi}{\partial r} \to 0\,, \quad \frac{\partial S}{\partial\theta} \to 0\,, \quad \text{as } \theta \to 0\,, \tag{2.29}$$

$$\frac{1}{\sin\theta}\frac{\partial\psi}{\partial r} \to 0\,, \quad \frac{\partial S}{\partial\theta} \to 0\,, \quad \text{as } \theta \to \pi\,. \tag{2.30}$$

The problem (2.23)–(2.30) can be put in dimensionless form by defining

$$r^* = k^{-1/2}r, \qquad \theta^* = \theta, \qquad t^* = \chi\,(\rho c)_m^{-1}k^{-1}t,$$

$$\psi^* = \rho_f C_f\,\chi^{-1}k^{-1/2}\psi\,, \qquad S^* = 4\pi\chi q^{-1}k^{1/2}S\,.$$

Omitting the asterisks to simplify the notation, we obtain the system

$$\frac{1}{r^2}\frac{\partial}{\partial\theta}\left(\frac{1}{\sin\theta}\frac{\partial\psi}{\partial\theta}\right) + \frac{1}{\sin\theta}\frac{\partial^2\psi}{\partial r^2} =$$

$$= Ra\left(\cos\theta\,\frac{\partial S}{\partial\theta} + r\sin\theta\,\frac{\partial S}{\partial r}\right), \tag{2.31}$$

$$\frac{\partial S}{\partial t} + \frac{1}{r^2\sin\theta}\frac{\partial(S,\psi)}{\partial(r,\theta)} =$$

$$\frac{1}{r^2\sin\theta}\left[\frac{\partial}{\partial r}\left(r^2\sin\theta\,\frac{\partial S}{\partial r}\right) + \frac{\partial}{\partial\theta}\left(\sin\theta\,\frac{\partial S}{\partial\theta}\right)\right], \tag{2.32}$$

$$\nabla\psi = 0\,, \quad S = 0, \quad t = 0\,, \tag{2.33}$$

$$\frac{1}{r}\,|\nabla\psi| \to 0\,, \quad S \to 0, \quad \text{as } r \to \infty\,, \tag{2.34}$$

$$\psi \to 0, \quad \frac{r^2}{2}\int_0^\pi \frac{\partial S}{\partial r}\sin\theta\, d\theta \to -1\,, \quad \text{as } r \to 0\,, \tag{2.35}$$

$$\frac{1}{\sin\theta}\frac{\partial\psi}{\partial r} \to 0\,, \quad \frac{\partial S}{\partial\theta} \to 0, \quad \text{as } \theta \to 0\,, \tag{2.36}$$

$$\frac{1}{\sin\theta}\frac{\partial\psi}{\partial r} \to 0, \quad \frac{\partial S}{\partial\theta} \to 0\,, \quad \text{as } \theta \to \pi\,, \tag{2.37}$$

where $Ra$ is the Rayleigh number defined by

$$Ra = \mu^{-1}\chi^{-2}k\rho_f^2 C_f\, g\alpha\, q/4\pi\,. \tag{2.38}$$

The problem is further simplified as the system (2.31)−(2.37) is invariant under the transformation

$$\tilde{r} = \lambda\, r, \quad \tilde{\theta} = \theta\,, \quad \tilde{t} = \lambda^2 t\,,$$

$$\tilde{\psi} = \lambda\psi\,, \quad \tilde{S} = \lambda^{-1}S\,,$$

for any $\lambda \in \mathbb{R}$.

This enables us to take the solutions of (2.31)−(2.37) in the form

$$\psi(r, \theta, t) = t^{1/2}F(\tau, \theta), \qquad S(r, \theta, t) = t^{-1/2}G(\tau, \theta)\,, \tag{2.39}$$

where the new variable $\tau$ is given by

$$\tau = \frac{1}{2}t^{-1/2}r\,. \tag{2.40}$$

Introducing (2.39) in (2.31)−(2.37) and making the change (2.40), we obtain the corresponding system for the new unknowns $F$ and $G$ as

$$\frac{1}{\tau^2}\frac{\partial}{\partial\theta}\left(\frac{1}{\sin\theta}\frac{\partial F}{\partial\theta}\right) + \frac{1}{\sin\theta}\frac{\partial^2 F}{\partial\tau^2} =$$

$$4\,Ra\left(\cos\theta\frac{\partial G}{\partial\theta} + \tau\sin\theta\frac{\partial G}{\partial\tau}\right), \tag{2.41}$$

$$\frac{1}{2}\frac{\partial(G, F)}{\partial(\tau, \theta)} = \frac{\partial}{\partial\tau}\left(\tau^2\sin\theta\frac{\partial G}{\partial\tau}\right) + \frac{\partial}{\partial\theta}\left(\sin\theta\frac{\partial G}{\partial\theta}\right) +$$

$$+\ 2\tau^2\sin\theta\left(G + \tau\frac{\partial G}{\partial\tau}\right), \tag{2.42}$$

$$\tau^{-1}F \to 0, \quad \tau G \to 0, \quad \text{as } \tau \to \infty\,, \tag{2.43}$$

$$F \to 0\,, \quad \tau^2\int_0^\pi \frac{\partial G}{\partial\tau}\sin\theta\, d\theta \to -1, \qquad \text{as } \tau \to 0\,, \tag{2.44}$$

$$\frac{1}{\sin\theta}\frac{\partial F}{\partial \tau} \to 0, \quad \frac{\partial G}{\partial \theta} \to 0, \quad \text{as } \theta \to 0\,, \tag{2.45}$$

$$\frac{1}{\sin\theta}\frac{\partial F}{\partial \tau} \to 0\,, \quad \frac{\partial G}{\partial \theta} \to 0\,, \quad \text{as } \theta \to \pi\,. \tag{2.46}$$

Equations (2.33) and (2.34) imply only one condition; i.e., (2.43).

The solution of the system (2.41)–(2.46) is obtained by means of a standard perturbation analysis which assumes power-series expansions with respect to $Ra$ for both functions $F$ and $G$:

$$F = F_0 + Ra\,F_1 + \cdots, \tag{2.47}$$

$$G = G_0 + Ra\,G_1 + \cdots\,. \tag{2.48}$$

The functions $F_i$ and $G_i$ are found by substituting (2.47) and (2.48) into (2.41)–(2.46) and solving the equations obtained by collecting together terms containing the same power of $Ra$.

Since at $Ra = 0$ there is no fluid motion ($\psi = 0$), we take

$$F_0(\tau, \theta) = 0\,. \tag{2.49}$$

Thus the zeroth-order function $G_0$ is given by

$$\frac{\partial}{\partial \tau}\left(\tau^2 \sin\theta \frac{\partial G_0}{\partial \tau}\right) + \frac{\partial}{\partial \theta}\left(\sin\theta \frac{\partial G_0}{\partial \theta}\right) + $$

$$+\ 2\tau^2 \sin\theta \left(G_0 + \tau\frac{\partial G_0}{\partial \tau}\right) = 0\,, \tag{2.50}$$

$$\tau G_0 \to 0, \quad \text{as } \tau \to \infty\,, \tag{2.51}$$

$$\tau^2 \int_0^{\pi} \frac{\partial G_0}{\partial \tau} \sin\theta \, d\theta \to -1, \quad \text{as } \tau \to 0\,, \tag{2.52}$$

$$\frac{\partial G_0}{\partial \theta} \to 0\,, \quad \text{as } \theta \to 0, \theta \to \pi\,. \tag{2.53}$$

It follows immediately that

$$G_0(\tau, \theta) = \frac{1}{2\tau}\,\mathrm{erfc}\,(\tau)\,, \tag{2.54}$$

where the error-functions are defined as usual:

$$\mathrm{erf}\,(\tau) = \frac{2}{\pi^{1/2}}\int_0^{\tau} \exp(-t^2)\,dt, \quad \mathrm{erfc}\,(\tau) = 1 - \mathrm{erf}\,(\tau)\,. \tag{2.55}$$

We can now obtain step by step the other functions $F_k$ and $G_k$. If we know $F_k$ and

$G_k$, $\forall k < n$ $(n \geqslant 1)$ then we find $F_n$ from

$$\frac{1}{\tau^2}\frac{\partial}{\partial\theta}\left(\frac{1}{\sin\theta}\frac{\partial F_n}{\partial\theta}\right) + \frac{1}{\sin\theta}\frac{\partial^2 F_n}{\partial\tau^2} = 4\left(\cos\theta\,\frac{\partial G_{n-1}}{\partial\theta} + \tau\sin\theta\,\frac{\partial G_{n-1}}{\partial\tau}\right), \tag{2.56}$$

$$\tau^{-1}F_n \to 0, \quad \text{as } \tau \to \infty, \tag{2.57}$$

$$F_n \to 0, \quad \text{as } \tau \to 0, \tag{2.58}$$

$$\frac{1}{\sin\theta}\frac{\partial F_n}{\partial\tau} \to 0, \quad \text{as } \theta \to 0, \theta \to \pi. \tag{2.59}$$

Similarly, we get $G_n$ from

$$\frac{\partial}{\partial\tau}\left(\tau^2\sin\theta\,\frac{\partial G_n}{\partial\tau}\right) + \frac{\partial}{\partial\theta}\left(\sin\theta\,\frac{\partial G_n}{\partial\theta}\right) +$$

$$+ 2\tau^2\sin\theta\left(G_n + \tau\frac{\partial G_n}{\partial\tau}\right) = \frac{1}{2}\sum_{k=0}^{n-1}\frac{\partial(G_k, F_{n-k})}{\partial(\tau,\theta)}, \tag{2.60}$$

$$\tau G_n \to 0, \quad \text{as } \tau \to \infty \tag{2.61}$$

$$\tau^2\int_0^\pi \frac{\partial G_n}{\partial\tau}\sin\theta\,d\theta \to 0, \quad \text{as } \tau \to 0, \tag{2.62}$$

$$\frac{\partial G_n}{\partial\theta} \to 0, \quad \text{as } \theta \to 0, \theta \to \pi. \tag{2.63}$$

Thus we find that

$$F_1(\tau,\theta) = \sin^2\theta f_1(\tau),$$

where

$$f_1(\tau) = \tau\,\mathrm{erfc}(\tau) + \mathrm{erf}(\tau)/2\tau - \exp(-\tau^2)/\pi^{1/2}, \tag{2.64}$$

$$G_1(\tau,\theta) = \cos\theta\, g_1(\tau), \tag{2.65}$$

where $g_1$ is defined by

$$\tau^2 g''_1 + 2(\tau^3+\tau)g'_1 + 2(\tau^2-1)g_1 = f_1\frac{\partial G_0}{\partial\tau}, \tag{2.66}$$

$$\tau g_1 \to 0, \quad \text{as } \tau \to \infty, \tag{2.67}$$

$$\tau^2 g'_1 \quad \text{finite as } \tau \to 0. \tag{2.68}$$

Let us remark that

$$\Phi_1(\tau) = \tau^{-2}\exp(-\tau^2) \tag{2.69}$$

satisfies the homogeneous equation (2.66). Looking for the second fundamental solution in the form

$$\Phi_2(\tau) = \Phi_1(\tau)\, h(\tau)\,,$$

we obtain

$$h''(\tau) = 2\,(\tau + \tau^{-1})\, h'(\tau)\,,$$

which yields

$$\Phi_2(\tau) = \tau^{-2} \exp(-\tau^2) \int_0^{\tau} t^2 \exp(t^2)\, \mathrm{d}t\,, \tag{2.70}$$

The solution of (2.66)–(2.68) is then obtained in the form

$$g_1(\tau) = C_1(\tau)\, \Phi_1(\tau) + C_2(\tau)\, \Phi_2(\tau)\,, \tag{2.71}$$

and the functions $C_1$ and $C_2$ are determined by the variation of arbitrary constants method. We then get

$$g_1(\tau) = \tau^{-2} \exp(-\tau^2) \int_0^{\tau} \varphi(t)\, t^2 \exp(t^2)\, \mathrm{d}t\,, \tag{2.72}$$

where

$$\varphi(t) = \int_t^{\infty} \left[ \frac{1}{2x} \operatorname{erfc}(x) + \frac{1}{\pi^{1/2}} \exp(-x^2) \right] \frac{f(x)}{x^3}\, \mathrm{d}x\,. \tag{2.73}$$

From $\psi_1$ we find that $\partial_\tau \psi_1 = 0$ is equivalent to

$$\tau^2 + \pi^{-1/2}\, \tau \exp(-\tau^2) = (\tau^2 + \frac{1}{2}) \operatorname{erf}(\tau)$$

which, besides $\tau = 0$, has only one solution: $\tau_{ex} = 0.88$; as $\partial_\theta \psi_1 = 0$ implies that $\theta \in \{0, \pi/2, \pi\}$ the extreme point of $\psi_1$ is located in $(\tau_{ex}, \pi/2)$. Hence, for sufficiently low Rayleigh numbers the streamline pattern at a given moment is as in Fig. 13. The corresponding temperature distribution is shown in Fig. 14.

For the steady state ($\partial_t = 0$) the computations are somewhat simpler. In this case the solution have the form

$$\psi(r, \theta) = rF(\cos\theta)\,, \qquad S(r, \theta) = r^{-1}\, G(\cos\theta)\,, \tag{2.74}$$

where the functions $F$, $G$:$[-1, 1] \to \mathbb{R}$ satisfy the following system, deduced from (2.31)–(2.37):

$$\frac{\mathrm{d}}{\mathrm{d}x}\left[ \frac{\mathrm{d}F}{\mathrm{d}x} + Ra\, x\, G(x) \right] = 0\,, \tag{2.75}$$

$$(1 - x^2) \frac{\mathrm{d}G}{\mathrm{d}x} = F(x)G(x)\,, \tag{2.76}$$

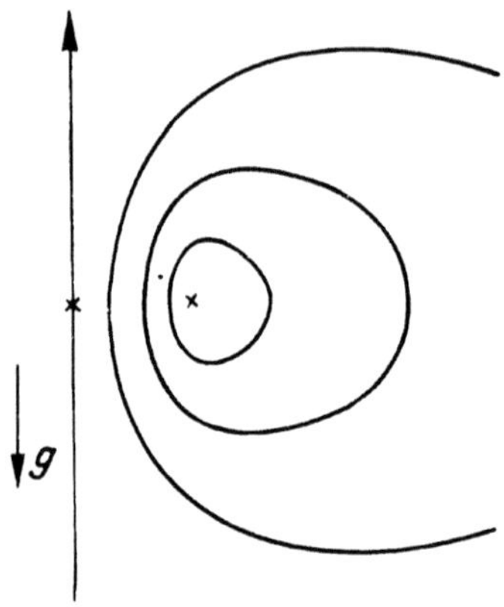


Fig. 13. Streamline pattern for fixed $t$.

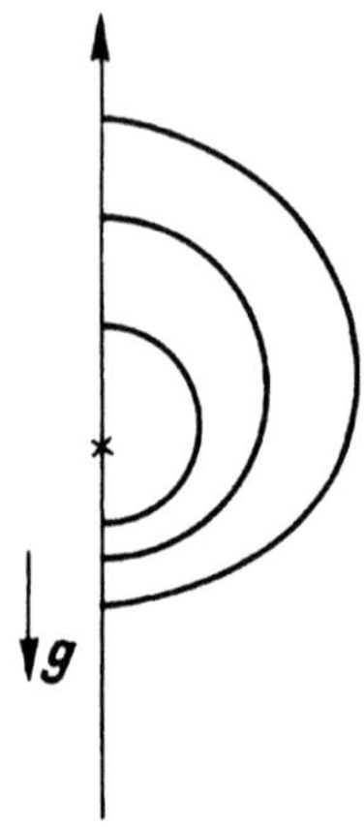


Fig. 14. Temperature distribution for fixed $t$.

$$\int_{-1}^{1} G(x)\,\mathrm{d}x = 2\,, \tag{2.77}$$

$$\lim_{x\to\pm 1} (1-x^2)^{-1/2} F(x) = 0\,. \tag{2.78}$$

Expanding again in powers of the Rayleigh number,

$$F(x) = \sum_{n\geq 0} F_n(x)\,Ra^n\,, \qquad G(x) = \sum_{n\geq 0} G_n(x)\,Ra^n\,, \tag{2.79}$$

we obtain in a straightforward manner

$$F_0(x) = 0\,, \qquad G_0(x) = 1\,, \tag{2.80}$$

$$F_1(x) = \frac{1}{2}(1-x^2), \qquad G_1(x) = \frac{1}{2}x\,, \tag{2.81}$$

$$F_2(x) = \frac{1}{2}(x-x^3), \qquad G_2(x) = \frac{5}{72}(3x^2-1)\,, \tag{2.82}$$

$$F_3(x) = \frac{5}{288}(1+2x^2-3x^4)\,, \qquad G_3(x) = \frac{1}{288}(5x+23x^3)\,, \tag{2.83}$$

and so on.

As with the heated sphere, we have two different flow patterns. The first is shown in Fig. 15, for $Ra = 1$, and the second in Fig. 16, for $Ra = 6$.

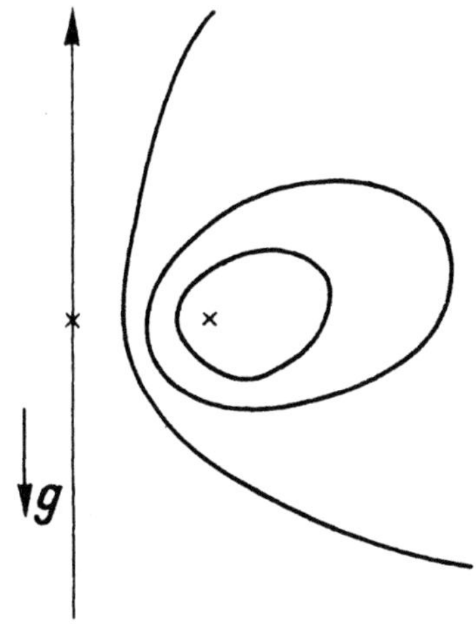


Fig. 15. Stationary streamlines for $Ra = 1$.

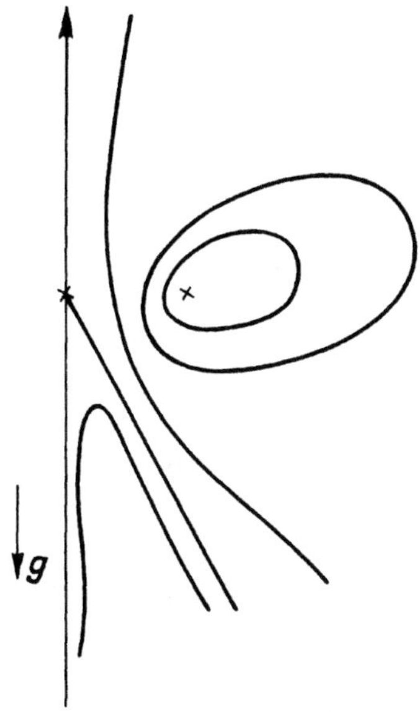


Fig. 16. Stationary streamlines for $Ra = 6$.

## 3. Natural Convection Between Two Horizontal Concentric Cylinders

The temperature distributions and streamline patterns presented in Section 2 of Chapter 2 and in Section 2 of this chapter are valid for sufficiently low Rayleigh numbers. We can also determine the range of Rayleigh number in which a series solution obtained by perturbation techniques and terminated after a number of iterations is a good approximation of the solution. But the two-cell pattern and, more importantly the three-cell patterns are out of such control.

The solution of the problem considered here is found without imposing any explicit or implicit restriction upon the Rayleigh number. And we were happy to find that the discussions and the patterns are similar to those mentioned above. The results given below, as well as being instrinsically interesting, can be considered as a test of validity for those control-less, multi-cell cases.

We consider the domain between two horizontal, concentric impervious cylinders of radii $R_1$ and $R_2$ ($R_1 < R_2$). Let O be the origin of the co-ordinate system $Ox_1x_2x_3$ such that $Ox_3$ is the axis of symmetry and the negative $Ox_1$ direction is the direction in which the gravitational force acts. The governing system of equations is, as usual (see (2.34)–(2.36) in Chapter 2),

$$\operatorname{div} \mathbf{u} = 0\,, \tag{3.1}$$

$$\frac{\mu}{k}\mathbf{u} + \nabla p = \rho_f\left[1 - \alpha\,(T - T_r)\right]\mathbf{g}\,, \tag{3.2}$$

$$\rho_f\, C_f\, \mathbf{u}\nabla T = \chi\Delta T, \tag{3.3}$$

for any $|\mathbf{x}| \in (R_1, R_2)$, $|\mathbf{x}| = (x_1^2 + x_2^2)^{1/2}$; it is to be solved subject to the boundary conditions

$$\mathbf{u}\cdot\mathbf{v} = 0, \quad |\mathbf{x}| = R_i, \quad i = 1, 2, \tag{3.4}$$

$$T = T_i, \quad |\mathbf{x}| = R_i, \quad i = 1, 2. \tag{3.5}$$

We put this problem in dimensionless form by defining

$$\mathbf{x}^* = R_1^{-1}\mathbf{x}, \qquad \varepsilon^* = R_1^{-1}(R_2 - R_1),$$

$$\mathbf{u}^* = \rho_f\, C_f\, R_1\, \chi^{-1}\mathbf{u}, \qquad T^* = (T_1 - T_2)^{-1}(T - T_2),$$

$$h^* = \mu^{-1}\chi^{-1}\rho_f C_f k\left[p + \rho_f g(1 + \alpha(T_r - T_2))x_1\right],$$

$$Ra^* = \mu^{-1}\chi^{-1}\rho_f^2 C_f (R_2 - R_1) k g\alpha (T_1 - T_2).$$

Omitting the asterisks to simplify the notation, we obtain

$$\operatorname{div}\mathbf{u} = 0, \qquad |\mathbf{x}| \in (1, 1 + \varepsilon), \tag{3.6}$$

$$\mathbf{u} + \nabla h = \varepsilon^{-1}\, Ra\, T\mathbf{e}_1, \qquad |\mathbf{x}| \in (1, 1 + \varepsilon), \tag{3.7}$$

$$\mathbf{u}\,\nabla T = \Delta T, \qquad |\mathbf{x}| \in (1, 1 + \varepsilon), \tag{3.8}$$

$$\mathbf{u}\cdot\mathbf{v} = 0, \qquad |\mathbf{x}| = 1, |\mathbf{x}| = 1 + \varepsilon, \tag{3.9}$$

$$T = 1, \qquad |\mathbf{x}| = 1, \tag{3.10}$$

$$T = 0, \qquad |\mathbf{x}| = 1 + \varepsilon. \tag{3.11}$$

REMARK 3.1. In general, the way in which equations are put into dimensionless form is arbitrary and not important. However, in most cases we have made assumptions about the magnitudes of the various parameters, and so the process is not random.

We now change to the cylindrical polar coordinate system $(r, \theta, z)$, with $u_z = 0$ and $\partial/\partial z = 0$, and introduce the Stokes stream function $\psi$ by

$$r u_r = \frac{\partial \psi}{\partial \theta}, \qquad u_\theta = -\frac{\partial \psi}{\partial r}. \tag{3.12}$$

Eliminating the pressure, the problem (3.6)–(3.11) reduces to

$$r^2\frac{\partial^2 T}{\partial r^2} + r\frac{\partial T}{\partial r} + \frac{\partial^2 T}{\partial \theta^2} = r\left(\frac{\partial T}{\partial r}\frac{\partial \psi}{\partial \theta} - \frac{\partial T}{\partial \theta}\frac{\partial \psi}{\partial r}\right), \tag{3.13}$$

$$r^2\frac{\partial^2 \psi}{\partial r^2} + r\frac{\partial \psi}{\partial r} + \frac{\partial^2 \psi}{\partial \theta^2} = \frac{rRa}{\varepsilon}\left(r\sin\theta\,\frac{\partial T}{\partial r} + \cos\theta\,\frac{\partial T}{\partial \theta}\right), \tag{3.14}$$

$$\psi = 0, \quad T = 1, \quad r = 1, \tag{3.15}$$

$$\psi = 0 , \quad T = 0 , \quad r = 1 + \varepsilon \ . \tag{3.16}$$

So far the analysis of the problem still seems to be complicated. We have simplified it by assuming that ε is sufficiently small for the perturbation method to hold; we now introduce a new variable, $s \in [0, 1]$, by

$$r = 1 + \varepsilon s \ . \tag{3.17}$$

Thus the system (3.13)−(3.16) becomes

$$A(\psi) = B(T) , \tag{3.18}$$

$$A(T) = (\varepsilon + \varepsilon^2 s)\left(\frac{\partial T}{\partial s}\frac{\partial \psi}{\partial \theta} - \frac{\partial T}{\partial \theta}\frac{\partial \psi}{\partial s}\right), \tag{3.19}$$

$$\psi = 0 , \quad T = 1 , \quad s = 0 , \tag{3.20}$$

$$\psi = 0 , \quad T = 0 , \quad s = 1 , \tag{3.21}$$

where the operator $A$ is defined by

$$A = A_0 + \varepsilon A_1 + \varepsilon^2 A_2 ,$$

$$A_0 = \frac{\partial^2}{\partial s^2} ,$$

$$A_1 = 2s\frac{\partial^2}{\partial s^2} + \frac{\partial}{\partial s} ,$$

$$A_2 = s^2\frac{\partial^2}{\partial s^2} + s\frac{\partial}{\partial s} + \frac{\partial^2}{\partial \theta^2} ,$$

and the operator $B$ by

$$B = B_0 + \varepsilon B_1 + \varepsilon^2 B_2 ,$$

$$B_0 = Ra \sin\theta \frac{\partial}{\partial s} ,$$

$$B_1 = Ra\left(2s \sin\theta \frac{\partial}{\partial s} + \cos\theta \frac{\partial}{\partial \theta}\right),$$

$$B_2 = Ra\left(s^2 \sin\theta \frac{\partial}{\partial s} + s\cos\theta \frac{\partial}{\partial \theta}\right) .$$

The peturbation method consists in expanding ψ and $T$ in power series in terms of ε:

$$\psi = \sum_{n \geqslant 0} \psi_n \varepsilon^n , \qquad T = \sum_{n \geqslant 0} T_n \varepsilon^n . \tag{3.22}$$

The functions $\psi_n$ and $T_n$ are found by substituting (3.22) into (3.18)−(3.21) and solving the equations obtained by identifying the coefficients of each power of $\varepsilon$.

The first problem to be solved is

$$A_0(T_0) = 0\,, \tag{3.23}$$

$$T_0 = 1\,, \quad s = 0\,, \tag{3.24}$$

$$T_0 = 0\,, \quad s = 1\,, \tag{3.25}$$

which yields

$$T_0 = 1 - s\,. \tag{3.26}$$

The problem for the zeroth-order function $\psi_0$ is then

$$A_0(\psi_0) = B_0(T_0)\,, \tag{3.27}$$

$$\psi_0 = 0\,, \quad s = 0\,, \quad s = 1\,. \tag{3.28}$$

It has one solution:

$$\psi_0 = \frac{1}{2} Ra\,(s - s^2) \sin\theta\,. \tag{3.29}$$

The next problem is

$$A_0(T_1) + A_1(T_0) = \frac{\partial T_0}{\partial s}\frac{\partial \psi_0}{\partial \theta}\,, \tag{3.30}$$

$$T_1 = 0\,, \quad s = 0, s = 1\,. \tag{3.31}$$

It follows that

$$T_1 = -\frac{1}{2}(s - s^2) + \frac{Ra}{24}(s - 2s^3 + s^4)\cos\theta\,. \tag{3.32}$$

Consequently $\psi_1$ is given by

$$A_0(\psi_1) + A_1(\psi_0) = B_0(T_1) + B_1(T_0)\,, \tag{3.33}$$

$$\psi_1 = 0\,, \quad s = 0\,, s = 1\,, \tag{3.34}$$

which yields directly

$$\psi_1 = \frac{Ra}{6}(s - 3s^2 + 2s^3)\sin\theta - \frac{Ra^2}{480}(2s - 5s^2 + 5s^4 - 2s^5)\sin 2\theta\,. \tag{3.35}$$

The first problem in second order is

$$A_0(T_2) + A_1(T_1) + A_2(T_0) = \frac{\partial(T_0, \psi_1)}{\partial(s, \theta)} + \frac{\partial(T_1, \psi_0)}{\partial(s, \theta)} + s\frac{\partial(T_0, \psi_0)}{\partial(s, \theta)}\,, \tag{3.36}$$

$$T_2 = 0\,, \quad s = 0\,, s = 1\,, \tag{3.37}$$

where $\partial(\,,\,)/\partial(\,,\,)$ denotes, as usual, the Jacobian. After some lengthy computations we obtain

$$T_2 = -\frac{(s-s^2)}{12}\left[4s + 1 + \frac{Ra^2}{3360}(60s^5 - 120s^4 + 18s^3 + \right.$$

$$+ 123s^2 - 17s - 17) + \frac{Ra}{60}(54s^3 - 66s^2 - 16s - 1)\cos\theta +$$

$$\left. + \frac{Ra^2}{480}(4s^5 - 10s^4 + 2s^3 + 7s^2 - s - 1)\cos 2\theta\right]. \tag{3.38}$$

The calculations become more and more laborious, and shall not be presented here beyond this approximation.

Nevertheless, this approximation yields an interesting formula for the Nusselt number, defined by

$$Nu = \int_0^{2\pi} \left.\frac{\partial T}{\partial s}\right|_{s=0} d\theta \Big/ \int_0^{2\pi} \left.\frac{\partial T_c}{\partial s}\right|_{s=0} d\theta, \tag{3.39}$$

where $T_c$ is the pure-conduction temperature distribution; hence $T_c$ satisfies the system (3.18)–(3.21) when $Ra = 0$; i.e.,

$$\psi_c = 0, \quad A(T_c) = 0,$$
$$T_c = 1, \quad s = 0,$$
$$T_c = 0, \quad s = 1,$$

which yields

$$T_c = 1 - \ln(1 + \varepsilon s)/\ln(1 + \varepsilon). \tag{3.40}$$

Taking into account that

$$\frac{\ln(1+\varepsilon)}{\varepsilon} = 1 - \frac{\varepsilon}{2} + \frac{\varepsilon^2}{3} + O(\varepsilon^3), \tag{3.41}$$

we finally obtain from (3.39)

$$Nu = 1 + \frac{17\, Ra^2}{40\,320}\varepsilon^2 + O(\varepsilon^3). \tag{3.42}$$

Also, in the second-order approximation we have

$$\psi = \frac{\varepsilon\, Ra\,(s - s^2)\sin\theta}{6}\left[\frac{3}{\varepsilon} + 1 - 2s - \frac{Ra\cos\theta}{40}(2s^3 - 3s^2 - 3s + 2)\right]. \tag{3.43}$$

We assume that $n = 3/\varepsilon > 1$; then, from (3.43) we find that there are three kinds of pattern for half of the configuration (the solution is symmetrical with respect to $Ox_1x_3$):

(a) If $Ra < 20(n - 1)$, then the pattern is unicellular. The right hand side of

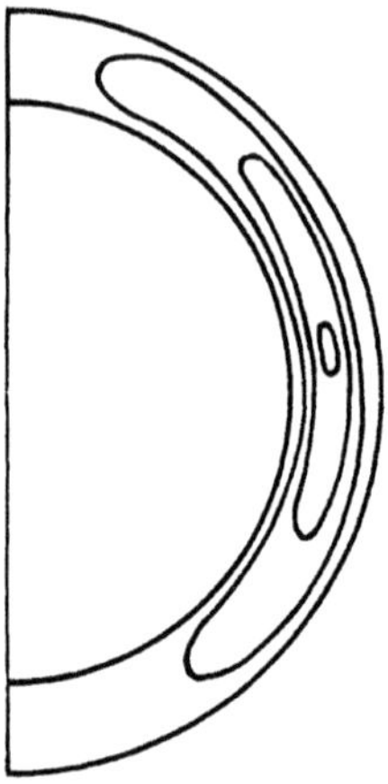

Fig. 17. Streamline pattern for $Ra = 80$.

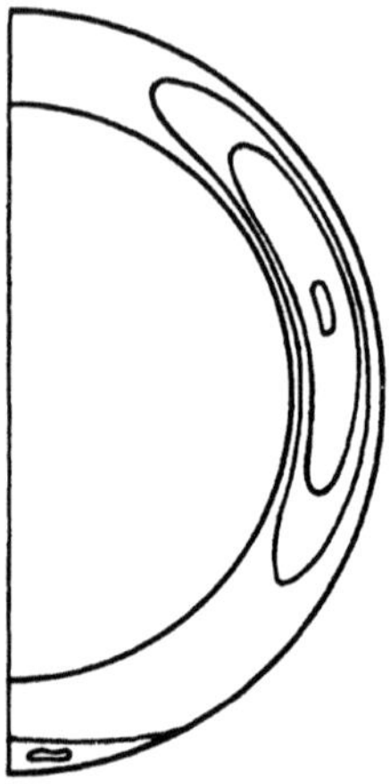

Fig. 18. Streamline pattern for $Ra = 180$.

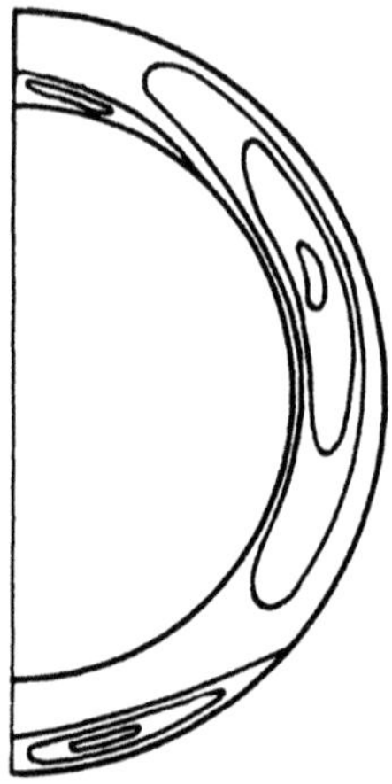

Fig. 19. Streamline pattern for $Ra = 296$.

Fig. 17 shows the streamlines for $n = 9$ and $Ra = 80$.

(b) If $20(n-1) < Ra < 20(n+1)$, then in the lower region of the enclosure there appears a second roll (Fig. 18, for $n = 9$ and $Ra = 180$). The pattern is quite stable as there is little buoyancy to induce the fluid to rise; it seems in fact that this is a stagnant portion.

(c) If $Ra > 20(n+1)$, a third small roll appears at the top of the inner cylinder (Fig. 19 for $n = 9$ and $Ra = 296$).

We argue that the oscillating three-dimensional convection observed in various reported experiments has its origin in this, probably unstable, third roll.

## 4. Thermal Boundary-Layer Approximation for Vertical Flow

Heat transfer from a surface embedded in a porous medium through which a liquid is flowing is of great practical importance in many branches of engineering. For example, convective flow in a porous medium is of considerable importance because of the use of geothermal energy for power generation.

The thermal boundary-layer approximation is based on the assumption that convection takes place in a thin layer around the heating surface. The analogy with the classical boundary-layer theory for a free viscous fluid is based on the fact that the motion in our case takes place at high Rayleigh numbers. Then, instead of Reynolds number, in the convection problem in porous media the parameter that indicates the possibility of this approach is the Rayleigh number.

In the previous sections of this chapter we have discussed problems in which the Rayleigh number takes small or moderate values, and the convection was represented by the complete system of equations derived in the first chapter (Section 4). For a high Rayleigh number this system of equations may be

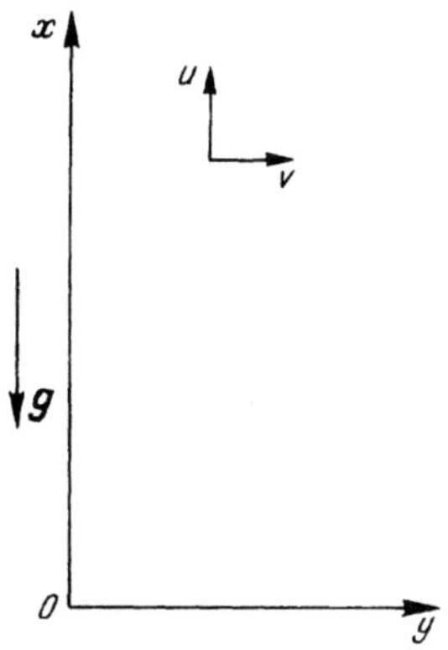


Fig. 20. Cartesian coordinate system.

simplified by a method similar to those proposed by Prandtl for the classical theory of a boundary layer in a free viscous fluid. We consider a rectangular Cartesian coordinate system with the origin fixed at the leading edge of the vertical surface such that the $x$-axis is directed upwards along the wall and the $y$-axis is normal to it (Fig. 20). The system of equations describing convective flow in a porous medium then becomes

$$\frac{\partial u}{\partial x} + \frac{\partial v}{\partial y} = 0\,, \tag{4.1}$$

$$u = -\frac{k}{\mu}\left(\frac{\partial p}{\partial x} + \rho g\right), \tag{4.2}$$

$$v = -\frac{k}{\mu}\frac{\partial p}{\partial y}, \tag{4.3}$$

$$\frac{\partial T}{\partial t} + \frac{1}{\sigma}\left(u\frac{\partial T}{\partial x} + v\frac{\partial T}{\partial y}\right) = \frac{\chi}{(\rho c)_m}\left(\frac{\partial^2 T}{\partial x^2} + \frac{\partial^2 T}{\partial y^2}\right), \tag{4.4}$$

$$\sigma = \frac{n\,\rho_f\,C_f + (1-n)\,\rho_s\,C_s}{\rho_f\,C_f}, \qquad (\rho c)_m = n\rho_f C_f + (1-n)\,\rho_s C_s\,, \tag{4.5}$$

$$\rho = \rho_\infty\,[1 - \alpha\,(T - T_\infty)]\,, \tag{4.6}$$

where $u$ and $v$ are Darcy velocities in the $x$ and $y$ directions, $\rho$ is the density, $\mu$ the viscosity, $\alpha$ the thermal expansion coefficient, $k$ the permeability of the saturated porous medium, $\lambda = \chi/\rho_f C_f$ the equivalent thermal diffusivity and $\chi$ the thermal conductivity of the saturated porous medium. The subscript $\infty$ denotes a value at infinity.

Eliminating $p$ from (4.2) and (4.3) by cross-differentiation, and using (4.6), we have

$$\frac{\partial v}{\partial x} - \frac{\partial u}{\partial y} + \frac{\alpha\rho_\infty kg}{\mu}\frac{\partial T}{\partial y} = 0\,. \tag{4.7}$$

If $\delta$ is the boundary-layer thickness, and if we suppose that $x \sim 1$, $y \sim \delta$, $u \sim 1$, $v \sim \delta$, $T \sim 1$, $\rho \sim 1$, $p \sim 1$ and $C \sim 1$, where $\sim$ denotes the order of magnitude, we find that $\frac{\partial}{\partial x} \sim 1$, $\frac{\partial}{\partial y} \sim \delta^{-1}$ and $\frac{\partial^2}{\partial y^2} \sim \delta^{-2}$. Consequently, from (4.1), (4.7) and (4.4) we obtain

$$\frac{\partial u}{\partial x} + \frac{\partial v}{\partial y} = 0\,, \tag{4.8}$$

$$\frac{\partial u}{\partial y} = \frac{\alpha\rho_\infty kg}{\mu}\frac{\partial T}{\partial y}\,, \tag{4.9}$$

$$\frac{\partial T}{\partial t} + \frac{1}{\sigma}\left(u\frac{\partial T}{\partial x} + v\frac{\partial T}{\partial y}\right) = \frac{\chi}{(\rho c)_m}\frac{\partial^2 T}{\partial y^2}\,, \tag{4.10}$$

which is the system of equations of thermal boundary-layer approximation in porous medium.

In the steady case, this system reduce to

$$\frac{\partial u}{\partial x} + \frac{\partial v}{\partial y} = 0\,, \tag{4.11}$$

$$\frac{\partial u}{\partial y} = \frac{\alpha\rho_\infty kg}{\mu}\frac{\partial T}{\partial y}\,, \tag{4.12}$$

$$u\frac{\partial T}{\partial x} + v\frac{\partial T}{\partial y} = \lambda\frac{\partial^2 T}{\partial y^2}\,. \tag{4.13}$$

If we introduce a characteristic length $L$, the Rayleigh number can be given as

$$Ra = \frac{\alpha\rho_\infty kg(T_w - T_\infty)L}{\mu\lambda}\,, \tag{4.14}$$

where $T_w$ is the wall temperature. By introducing the dimensionless variables and functions

$$X = \frac{x}{L}\,, \qquad Y = \frac{yRa^{1/2}}{L}\,, \qquad \tau = \frac{\lambda Ra}{\sigma L^2}t\,, \tag{4.15}$$

$$U = \frac{Lu}{\lambda Ra}\,, \qquad V = \frac{Lv}{\lambda Ra^{1/2}}\,, \qquad \theta = \frac{T - T_\infty}{T_w - T_\infty}\,, \tag{4.16}$$

we obtain the system of equations

$$\frac{\partial U}{\partial X} + \frac{\partial V}{\partial Y} = 0\,, \tag{4.17}$$

$$\frac{\partial U}{\partial Y} = \frac{\partial \theta}{\partial Y}, \tag{4.18}$$

$$\frac{\partial \theta}{\partial \tau} + U\frac{\partial \theta}{\partial X} + V\frac{\partial \theta}{\partial Y} = \frac{\partial^2 \theta}{\partial Y}. \tag{4.19}$$

Note that $\theta$ always must vanish at infinity and $V$ must be zero at $Y = 0$. We may then write Equation (4.19) in integral form: we integrate with respect to $Y$, using the divergent form and (4.17), and obtain

$$\frac{\partial}{\partial \tau}\int_0^\infty \theta \, \mathrm{d}Y + \frac{\partial}{\partial X}\int_0^\infty U\theta \, \mathrm{d}Y = -\left.\frac{\partial \theta}{\partial Y}\right|_{Y=0}. \tag{4.20}$$

This integral form of the transient energy equation will be very useful.

Of course, it is possible to introduce the stream function $\psi$ by using the classical relations

$$U = \frac{\partial \psi}{\partial Y}, \qquad V = -\frac{\partial \psi}{\partial X}, \tag{4.21}$$

and then we obtain the equations

$$\frac{\partial \psi}{\partial Y} = \theta, \tag{4.22}$$

$$\frac{\partial \theta}{\partial \tau} + \frac{\partial \psi}{\partial Y}\frac{\partial \theta}{\partial X} - \frac{\partial \psi}{\partial X}\frac{\partial \theta}{\partial Y} = \frac{\partial^2 \theta}{\partial Y^2} \tag{4.23}$$

Equation (4.22) indicates that the vertical velocity and temperature distribution are of the same shape. This equation may be used to show that at the wall the vertical velocity varies in the same way as the prescribed wall temperature. In particular, for an isothermal wall, vertical velocity is constant along the wall. This conclusion is a consequence of the conditions imposed at infinity: the vanishing of the vertical velocity and of the temperature field, in dimensionless form.

We now examine the phenomena arising in the thermal boundary layer, by taking some typical examples.

## 4.1. STEADY STATE SOLUTIONS

In the steady state, it is possible to obtain similar solutions in some particular cases of interest.

We consider first an unbounded porous medium in a gravitational field, saturated with a fluid at temperature $T_\infty$ at rest containing a longitudinal line heat source. In a plane normal to its length, let the trace of the source be the origin of the Cartesian coordinates $X$ and $Y$, with $X$ in the vertical direction. We make the transformations

$$\eta = \frac{Y}{X^{2/3}}, \tag{4.24}$$

$$\psi = X^{1/3} f(\eta), \tag{4.25}$$

$$\theta = X^{-1/3}\,\theta(\eta). \tag{4.26}$$

We have to solve Equations (4.22) and (4.23) (in the steady case) with the boundary conditions

$$f(0) = 0, \qquad f''(0) = 0, \qquad f'(+\infty) = 0, \qquad \theta(+\infty) = 0. \tag{4.27}$$

Of course, symmetry and the requirements at infinity have been taken into account.

From (4.22) and (4.23) we have

$$f' = \theta, \tag{4.28}$$

$$-(f\theta)' = 3\theta''. \tag{4.29}$$

Note that the condition $f''(0) = 0$ in (4.27) also implies that $\theta'(0) = 0$, as a consequence of the fact that (4.18) yields $f'' = \theta'$. Equation (4.28) is a direct consequence of the infinity conditions. A first integration of (4.29) produces

$$-f\theta = 3\theta', \tag{4.30}$$

the constants of integration being zero by virtue of the conditions (4.27). It follows that

$$-ff' = 3f'', \tag{4.31}$$

integration of which produces

$$6f' = A^2 - f^2, \tag{4.32}$$

$A^2$ being a constant of integration. Let $f = AF$. Then (4.32) can be integrated directly to give

$$F(\eta) = \tanh \frac{A}{6}\,\eta. \tag{4.33}$$

The constant $A$ is determined from (4.20), which in this case is of the form

$$\int_0^\infty f'\,\theta\,d\eta = \text{const.} \tag{4.34}$$

The constant is the convective part of the strength of the source. The flow pattern can be obtained by plotting curves of constant values of $\psi$ (Fig. 21), and the isotherms by using (4.25) (Fig. 22).

Another example is provided by a heated flat plate. In this case we use the transformations

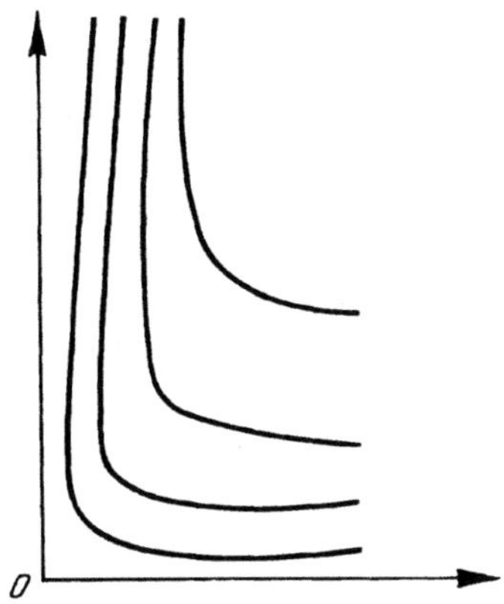


Fig. 21. Flow pattern.

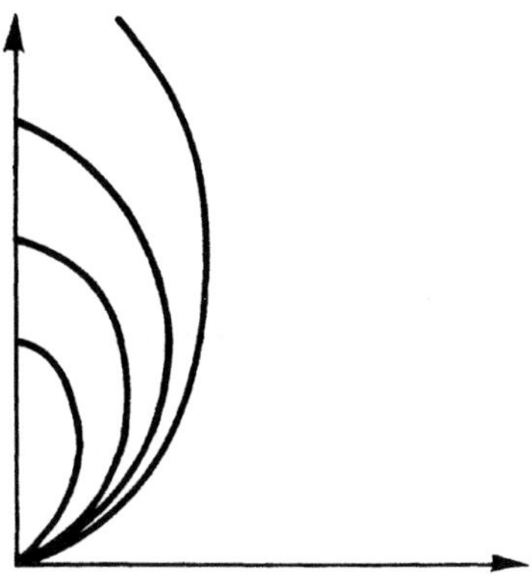

Fig. 22. Isotherms.

$$\eta = \frac{Y}{X^{1/2}} , \tag{4.35}$$

$$\psi = X^{1/2} f(\eta) . \tag{4.36}$$

From (4.22) and (4.23) we have

$$f' = \theta , \tag{4.37}$$

$$\theta'' = -\frac{1}{2}\,\theta' f, \tag{4.38}$$

and the boundary conditions are $f(0) = 0$, $\theta(0) = 1$, $f'(\infty) = 0$ and $\theta(\infty) = 0$. From (4.37) and (4.38) we obtain only one equation for $f$:

$$f''' + \frac{1}{2} ff'' = 0 . \tag{4.39}$$

This equation was integrated numerically by Cheng and Minkowycz (1977), who used the Runge – Kutta method. Figure 23 shows $\theta$ (or $f$) as function of $\eta$;

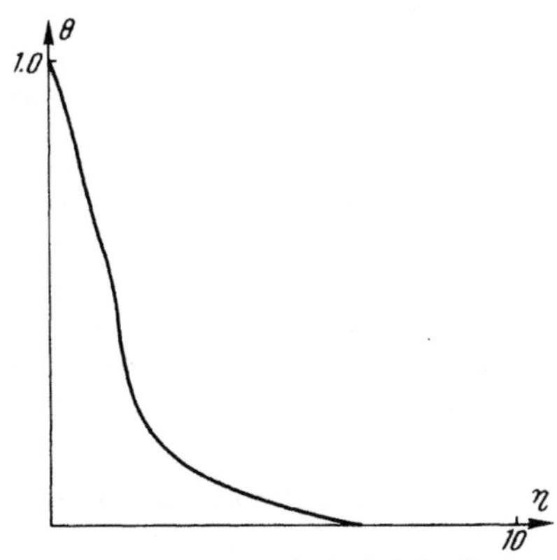


Fig. 23. Temperature (velocity) distribution.

the dimensionless temperature (or dimensionless vertical velocity) is seen to have a maximum equal to 1 at $\eta = 0$, and to decrease as $\eta$ is increased.

As is usual in boundary-layer theory, it is possible to define a local Rayleigh number instead of using (4.14), by the relation

$$Ra_x = \frac{\alpha \rho_\infty \, kg(T_w - T_\infty)\, x}{\mu\lambda} \,. \tag{4.40}$$

With this local Rayleigh number, the dimensionless similarity variables are

$$\eta = (Ra_x)^{1/2}(y/x)\,, \tag{4.41}$$

$$\psi = \lambda\,(Ra_x)^{1/2} f(\eta)\,, \tag{4.42}$$

$$\theta(\eta) = \frac{T - T_\infty}{T_w - T_\infty} \,. \tag{4.43}$$

Equations (4.37) and (4.38) remain the same, and we may obtain the equation for the boundary-layer thickness $\delta$ (the thickness of the hot-water zone). We define the edge of the boundary layer as that point where $\delta$ has a value of 0.01. Let $\eta_\delta$ be the value of $\eta$ at this point; it follows that

$$\frac{\delta}{x} = \frac{\eta_\delta}{(Ra_x)^{1/2}} \,. \tag{4.44}$$

The value of $\eta_\delta$ is 6.31.

The local heat-transfer rate along the surface of the flat plate is defined by $q = -\chi \left(\frac{\partial T}{\partial y}\right)_{y=0}$. Using this definition, the local Nusselt number, defined by

$$Nu_x = \frac{qx}{\chi\,(T_w - T_\infty)}\,,$$

is given by

$$\frac{Nu_x}{(Ra_x)^{1/2}} = -\,\theta'(0) = 0.4440\,. \tag{4.45}$$

These results were applied to the problem of free convection about a dike, using the values $T_w = 200°C$, $T_\infty = 15°C$, $k = 10^{-12}$ m$^2$, $\rho_\infty = 10^6$ g m$^{-3}$, $L = 300$ m, $\mu = 0.27$ g s$^{-1}$ m$^{-1}$, $\lambda = 6.3 \times 10^{-7}$ m$^2$ s$^{-1}$, $\alpha = 1.8 \times 10^{-4}$ K$^{-1}$ and $\chi = 0.58$ cal s$^{-1}$ K$^{-1}$ m$^{-1}$. Figure 24 shows the shape of the dimensional vertical velocity and temperature distribution about the dike. The velocity along the dike is constant at 0.0044 cm h$^{-1}$. The boundary layer thickness increases from zero at the origin to 80 m at 300 m, where the Rayleigh number is 575. If we use a permeability of $10^{-10}$ m$^{-2}$, the thickness of the boundary-layer at 300 m is only 8 m, and the Rayleigh number is 5750. In this case the slip velocity on

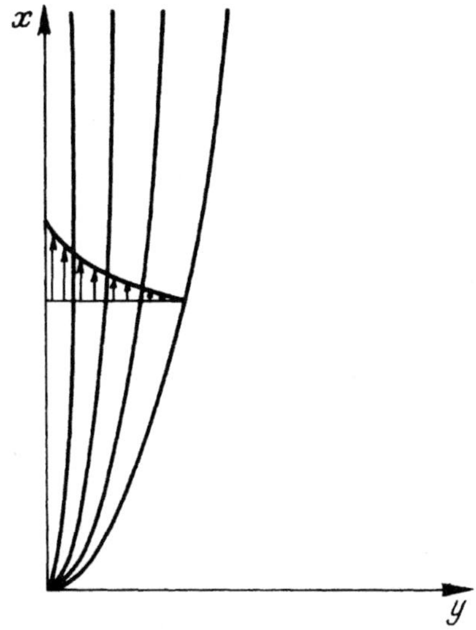


Fig. 24. Isotherms and velocity profiles.

the dike is 0.44 m $h^{-1}$. The results of this computation show that the hot-water zone is very thin for large Rayleigh numbers.

More interesting results are obtained if the impermeable vertical wall embedded in the porous medium is assumed to have a non-uniform temperature or if a non-uniform heat flux is assumed. We define a Rayleigh number in terms of a reference temperature $T_r$,

$$Ra = \frac{\alpha \rho_\infty k g (T_r - T_\infty) L}{\mu \lambda}, \tag{4.46}$$

and take the same dimensionless variables and functions as in (4.15) and (4.16), with a modification to take account of the non-uniformity of the temperature on the wall,

$$\theta = \frac{T - T_\infty}{T_w(X) - T_\infty}. \tag{4.47}$$

We introduce a function, defined on the wall:

$$S_w(X) = \frac{T_w(X) - T_\infty}{T_r - T_\infty}. \tag{4.48}$$

Then instead of (4.18) and (4.19), we have

$$\frac{\partial U}{\partial Y} = S_w(X) \frac{\partial \theta}{\partial Y}, \tag{4.49}$$

$$U\left(\frac{\partial \theta}{\partial X} + \theta \frac{\partial \ln S_w}{\partial X}\right) + V\frac{\partial \theta}{\partial Y} = \frac{\partial^2 \theta}{\partial Y^2}, \tag{4.50}$$

subject to the boundary conditions

$$V = 0, \quad \theta = 1, \quad Y = 0, \tag{4.51}$$

$$U = 0, \quad \theta = 0, \quad Y = \infty. \tag{4.52}$$

Introducing the steam function $\psi(X, Y)$,

$$U = \frac{\partial \psi}{\partial Y}, \qquad V = -\frac{\partial \psi}{\partial X},$$

we look for a new transformation:

$$\xi = X, \qquad \eta = S_w^{1/2} \frac{Y}{X^{1/2}}, \tag{4.53}$$

$$\psi = \xi^{1/2} S_w^{1/2}(\xi) f(\xi, \eta), \qquad \theta = \theta(\xi, \eta). \tag{4.54}$$

The resulting equations are

$$f'' = \theta', \tag{4.55}$$

$$\theta'' + \frac{B(\xi) + 1}{2} f\theta' - B(\xi) f' g = \left(f' \frac{\partial \theta}{\partial \xi} - \theta' \frac{\partial f}{\partial \xi}\right), \tag{4.56}$$

with the boundary conditions

$$f(\xi, 0) = 0, \quad \theta(\xi, 0) = 1, \tag{4.57}$$

$$f'(\xi, \infty) = 0, \quad \theta(\xi, \infty) = 0. \tag{4.58}$$

Here the primes indicate differentiation with respect to $\eta$, and $B(\xi)$ is defined by

$$B(\xi) = \xi S_w^{-1}(\xi) \frac{dS_w(\xi)}{d\xi}.$$

Integration of (4.55) with (4.58) yields

$$f' = \theta, \tag{4.59}$$

with conclusions similar to those drawn from (4.22). It is then clear that Equation (4.56) takes the final form

$$f''' + \frac{B(\xi) + 1}{2} ff'' - B(\xi)(f')^2 = \left(f' \frac{\partial f'}{\partial \xi} - f'' \frac{\partial f}{\partial \xi}\right), \tag{4.60}$$

and, from (4.57) and (4.58),

$$f(\xi, 0) = 0, \qquad f'(\xi, 0) = 1, \qquad f'(\xi, \infty) = 0. \tag{4.61}$$

As in the previous case, the local Nusselt number is given by

$$\frac{Nu_x}{Ra_x^{1/2}} = -\theta'(\xi, 0). \tag{4.62}$$

Na and Pop (1983) considered three cases:

$$S_w(\xi) = a\xi^{\alpha}, \tag{4.63}$$

$$S_w(\xi) = 1 + \xi^{n}, \tag{4.64}$$

$$S_w(\xi) = e^{n\xi}, \tag{4.65}$$

which they solved by finite-difference method. The numerical results may be found in the reference; we give only some of the values here. For (4.64) with $n = 1$, at $\xi = 0.99$, $\frac{Nu_x}{Ra_x^{1/2}} = 0.7315$, and at $\xi = 6.91$, $\frac{Nu_x}{Ra_x^{1/2}} = 0.9357$; for (4.65) with $n = 1$, at $\xi = 0.99$, $\frac{Nu_x}{Ra_x^{1/2}} = 0.9279$, and at $\xi = 6.91$, $\frac{Nu_x}{Ra_x^{1/2}} =$ $= 2.3864$. The case (4.63) is very interesting because it corresponds to $T_w(x) = T_\infty + bx^{\alpha}$, and (4.60) reduces to

$$f''' + \frac{\alpha + 1}{2} ff'' - \alpha (f')^2 = 0,$$

which is the case treated by Cheng and Minkowycz (1977).

If at the wall we have a non-uniform heat flux $q_w(x)$, instead of (4.46) we define a Rayleigh number

$$Ra_* = \frac{\alpha \rho_\infty k g q_w(x) L^2}{\lambda \mu \chi}, \tag{4.66}$$

and the appropriate dimensionless variables and functions:

$$X = \frac{x}{L}, \qquad Y = \frac{y Ra_*^{1/2}}{L}, \tag{4.67}$$

$$\theta = \frac{T - T_\infty}{\theta_w(X)}, \qquad \theta_w(X) = \frac{q_w(X) L}{\chi Ra_*^{1/3}} = C q_w^{2/3}(X), \tag{4.68}$$

$$\psi = Ra_*^{-1/3} X^{2/3} f(\xi, \eta), \qquad \theta = X^{2/3} g(\xi, \eta), \tag{4.69}$$

$$\xi = X, \qquad \eta = \frac{Y}{X^{1/3}}. \tag{4.70}$$

The boundary conditions are

$$V = 0, \quad \frac{\partial \theta}{\partial Y} = -1, \quad Y = 0, \tag{4.71}$$

$$V = 0, \quad \theta = 0, \quad Y = \infty. \tag{4.72}$$

Now, instead of (4.60) we obtain

$$f''' + \left[\frac{2}{3} + \frac{A(\xi)}{2}\right] ff'' - \left[\frac{1}{3} + A(\xi)\right] (f')^2 = \xi \left(f' \frac{\partial f'}{\partial \xi} - f'' \frac{\partial f}{\partial \xi}\right), \tag{4.73}$$

$$f(\xi, 0) = 0, \qquad f''(\xi, 0) = -1, \qquad f'(\xi, \infty) = 0, \tag{4.74}$$

$$A(\xi) = \frac{\xi}{\theta_w(\xi)} \frac{d\theta_w(\xi)}{d\xi} = \frac{2C}{3} \frac{\xi}{q_w(\xi)} \frac{dq_w(\xi)}{d\xi}. \tag{4.75}$$

There are only two cases for which similar solutions may be obtained: the case of constant heat flux on the wall ($q_w$ = const.) and the case of a heat flux proportional to $\xi^m$ ($q_w(\xi) = C_1\xi^\alpha$). The equations to be solved for these two cases are, respectively,

$$f''' + \frac{2}{3} ff'' - \frac{1}{3} (f')^2 = 0, \tag{4.76}$$

$$f''' + (\frac{2}{3} + \frac{1}{2}\alpha) ff'' - (\frac{1}{3} + \frac{2}{3}\alpha)(f')^2 = 0. \tag{4.77}$$

Numerical solutions for other problems of this type may be obtained.

### 4.2. NON-STEADY BOUNDARY LAYER

The problem of transient free convection in a porous medium adjacent to a vertical, semi-infinite flat plate with a step increase in wall temperature or surface heat flux, gives rise, as in the classical problem of a viscous boundary layer in a free fluid, to the singularity problem in passing from the initial stage when the leading edge effect is not felt to the steady state defined for large time. For small values of time the solutions for velocity and temperature are independent of $X$; for large values of time the solutions are independent of time. The singularity value of time depends on the vertical distance $X$. The heat-transfer characteristics change suddenly from transient, one-dimensional heat conduction to steady two-dimensional free convection.

In order to formulate the problem we consider a semi-infinite, saturated porous medium bounded by a vertical flat plate, both of which are initially at a uniform temperature $T_\infty$. Transient free convection in the porous medium adjacent to the plate will occur as a result of a step increase in wall temperature $T_w$ or surface heat flux $q_w$. We then take the boundary layer equations (4.20) and (4.22):

$$U = \theta, \tag{4.78}$$

$$\frac{\partial}{\partial \tau}\int_0^\infty \theta \, dY + \frac{\partial}{\partial X}\int_0^\infty U\theta \, dY = -\left.\frac{\partial \theta}{\partial Y}\right|_{Y=0}. \tag{4.79}$$

Substituting Equation (4.78) into Equation (4.79) yields

$$\frac{\partial}{\partial \tau}\int_0^\infty \theta Y + \frac{\partial}{\partial X}\int_0^\infty \theta^2 \, dY = -\left.\frac{\partial \theta}{\partial Y}\right|_{Y=0}, \tag{4.80}$$

which is to be solved subject to the intial condition

$$\theta(X, Y, 0) = 0 \tag{4.81}$$

and the boundary conditions

$$\theta(X, 0, \tau) = 1, \tag{4.82a}$$

$$-\frac{\partial \theta}{\partial Y}(X, 0, \tau) = Q_w, \tag{4.82b}$$

$$\theta(X, \infty, \tau) = 0. \tag{4.83}$$

Equations (4.82) represent the boundary conditions for a step increase in wall temperature and wall heat flux.

With the boundary condition (4.82a) we assume a temperature distribution of the form

$$\theta = \operatorname{erfc} \eta, \tag{4.84}$$

where erfc is the complementary error function

$$\operatorname{erfc} \eta = 1 - \operatorname{erf} \eta = 1 - \frac{2}{\sqrt{\pi}}\int_0^\infty e^{-\xi^2} \, d\xi,$$

and $\eta = Ra^{-1/2}\dfrac{Y}{H(X,\tau)}$, $H = \dfrac{\delta}{L}$, $\delta$ being the boundary-layer thickness. The form (4.84) satisfies the boundary conditions (4.82a) and (4.83). Substituting (4.84) into (4.80) after integrating yields

$$\frac{\partial H}{\partial \tau} + a\frac{\partial H}{\partial X} = \frac{2}{H} Ra^{-1}, \tag{4.85}$$

where $a = 2 - \sqrt{2} = 0.5857$, and we have used the values $\int_0^\infty \operatorname{erfc} \eta \, d\eta = \dfrac{1}{\sqrt{\pi}}$, $\int_0^\infty (\operatorname{erfc} \eta)^2 \, d\eta = \dfrac{(2-\sqrt{2})}{\sqrt{\pi}}$ and $\theta'(0) = \dfrac{2}{\sqrt{\pi}}$. We have to integrate (4.85) with the initial condition $H(X, 0) = 0$ and the boundary condition $H(0, \tau) = 0$; that is, we determine $H$ (or $\delta$) for a step increase in wall temperature. After $\delta$ is obtained, the surface heat flux is given by

$$q_w = \frac{2\chi(T_w - T_\infty)}{\delta\sqrt{\pi}}. \tag{4.86}$$

Note that $Q_w$ from (4.82b) is related to $q_w$ by

$$Q_w = \frac{L}{\chi\,(T_w - T_\infty)\,Ra^{1/2}}\,q_w = \frac{2L}{Ra^{1/2}\,\delta\sqrt{\pi}}$$

The solutions of (4.85) can be obtained by the method of characteristics. The differential system equivalent to (4.85) is

$$\frac{dX}{a} = d\tau = Ra\,\frac{H\,dH}{2}\,, \tag{4.87}$$

which has the characteristics

$$dX = a\,d\tau\,. \tag{4.88}$$

On each characteristic, $H$ is related by

$$Ra\,H dH = 2\,d\tau\;, \tag{4.89}$$

or

$$Ra\,aH\,dH = 2\,dX, \tag{4.90}$$

depending on whether the characteristic intercepts the $\tau$ – or the $X$-axis. Integrating Equation (4.89) with the initial condition $H(X, 0) = 0$ gives

$$H = 2\sqrt{\frac{\tau}{Ra}}\,, \tag{4.91}$$

and (4.90) with the boundary condition $H(0, \tau) = 0$ gives

$$H = 2\sqrt{\frac{X}{a\,Ra}} = 2.61\sqrt{\frac{X}{Ra}}\,. \tag{4.92}$$

The expression for $H$ changes from equation (4.91) to equation (4.92) along the limiting line characteristic

$$\tau = \frac{X}{a} = \frac{X}{0.5857}\,. \tag{4.93}$$

Equation (4.93) is a straight line which divides the $(X, \tau)$ plane into two regions: a lower region, with $H$ given by (4.91), and an upper region with $H$ given by (4.92). In the physical variables, the limit time obtained directly from (4.93) is given by

$$t_c = \frac{\sigma L}{a\lambda}\frac{x}{Ra} = \frac{\sigma}{\lambda a}\frac{x^2}{Ra_x}\,. \tag{4.94}$$

Then for $t < t_c$, we have

$$\theta = \operatorname{erfc}\left(\frac{y}{2}\sqrt{\frac{\sigma}{\lambda t}}\right), \tag{4.95a}$$

$$q_w = \chi\,(T_w - T_\infty)\left|\sqrt{\frac{\sigma}{\lambda\pi t}}\right., \tag{4.95b}$$

and for $t > t_c$

$$\theta = \operatorname{erfc}\left(\frac{y}{2x}\sqrt{a\,Ra_x}\right), \qquad q_w = \frac{\chi\,(T_w - T)}{x}\sqrt{\frac{a\,Ra_x}{\pi}}. \tag{4.96}$$

From the definition of the local Nusselt number

$$Nu_x = \frac{q_w x}{\chi\,(T_w - T_\infty)},$$

we have

$$\frac{Nu_x}{\sqrt{Ra_x}} = 0.4318. \tag{4.97}$$

Note that Equations (4.95) are independent of $x$, and that the solution represents the transient heat conduction in a semi-infinite porous medium. Equations (4.96), however, are independent of $t$, and their solution represents steady-state free convection. In this problem $t_c$ represents the time required to reach the steady state. It is interesting to observe the good agreement between (4.45) and (4.97).

It is known that in general we have the maximum vertical velocity along the plate, and according to (4.78) this is given by (4.95a). We can then define the limiting extent of the leading-edge effect as

$$x_{max} = \int_0^t u\,(x, 0, t)\,\mathrm{d}t = k\rho_\infty\alpha g\,(T_w - T_\infty)\,\frac{t}{\mu} = \frac{\lambda}{L}\,Ra\;t.$$

The problem with the boundary condition (4.82b) may be treated in similar manner. We look for a solution of the form

$$\theta = \frac{Q_w\,H\sqrt{\pi\,Ra}}{2}\operatorname{erfc}\eta\,, \tag{4.98}$$

which satisfies (4.82b) and (4.83). From (4.80) the equation for the boundary-layer thickness is

$$\frac{\partial H}{\partial \tau} + bH\frac{\partial H}{\partial X} = \frac{1}{H\,Ra}, \tag{4.99}$$

where $b = \frac{3}{4}(2 - \sqrt{2})\,Q_w\sqrt{\pi\,Ra}$. Equation (4.99) with the boundary condition $H(0, \tau) = 0$ and the initial condition $H(X, 0) = 0$ gives the characteristic line

$$\frac{\mathrm{d}\tau}{\mathrm{d}X} = \frac{1}{bH}. \tag{4.100}$$

On each characteristic, $H$ is given by

$$H = \sqrt{\frac{2\tau}{Ra}}, \quad t < t_c, \tag{4.101}$$

or

$$H = \left(\frac{3X}{b\,Ra}\right)^{1/3}, \quad t > t_c. \tag{4.102}$$

The critical value of time is obtained directly from (4.100), in terms of physical variables, as

$$t_c = \frac{\sigma}{\lambda}\left(\frac{2L}{\pi}\right)^{1/3} \frac{1}{[Q_w(2-\sqrt{2})]^{2/3}} \frac{x^{5/3}}{Ra_x}. \tag{4.103}$$

Equation (4.103) yields a curve which separates the plane $(x, t)$ into two regions: a lower region with $H$ given by (4.101) when the solutions are independent of $x$, and an upper region with $H$ given by (4.102) when the solutions are independent of $t$. As in the previous case, we have a solution for transient heat conduction in a semi-infinite plane with a step increase in surface heat flux which is in good agreement with the temperature obtained from (4.101) and (4.98). The solution (4.102) represents steady heat convection on the plate, and it is also in good agreement with the exact solution.

Figures 25 and 26 show $H$ as a function of $\tau$ at different values of $X$ for, respectively, a step increase in wall temperature and a step increase in heat flux. From these figures it is clear that at $t = t_c$ there is a singularity in the solution.

It is also possible to imagine approximate solutions for Equation (4.85) or (4.99), and in this case the discontinuity disappears because it was used an analytical function in $X$ and $\tau$ (Cheng and Pop (1984)).

Physically, this change in characteristics can be explained in terms of the time required for the propagation of the leading-edge effect.

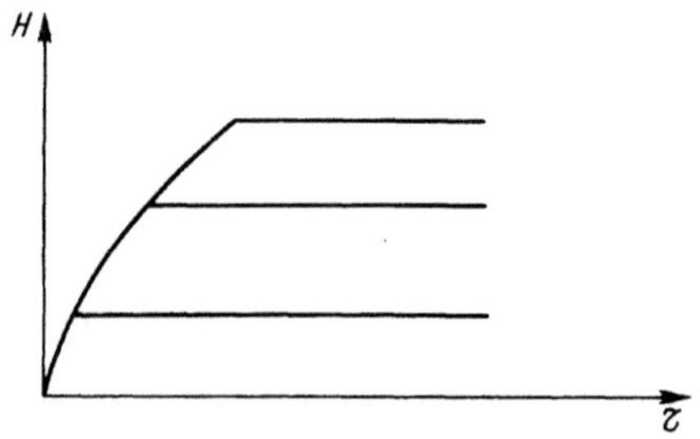


Fig. 25. Boundary-layer thickness for different $x$, for step increases in wall temperature.

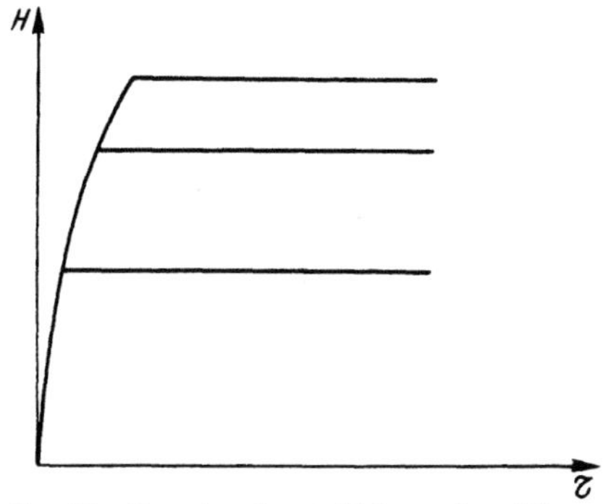


Fig. 26. Boundary-layer thickness for different $x$, for step increases in heat flux.

## 5. Comments and Bibliographical Notes

A non-linear analysis of natural convection in a horizontal porous layer of infinite extent has been presented in Section 1. The non-linear equations describing the velocity field and the temperature distribution are solved by a successive-approximation method suggested independently by Malkus and Veronis (1958) and Gor'kov (1958), in which the small parameter is a measure of the deviation from the static state. The stability of these finite-amplitude steady solutions with respect to infinitesimally small disturbances is examined by the procedure used by Schlüter *et al.* (1965); i.e., the stability equations are studied by a perturbation technique similar to the Gor'kov-Malkus-Veronis method.

The conditions (1.26) and (1.27) were first used by Lindstedt (1883) to avoid such a 'hidden' parameter indeterminacy which arose in some problems in celestial mechanics.

Natural convection about a heated sphere in an unbounded region of a porous medium was studied for the first time by Yamamoto (1974). In Section 2 we have followed his procedure, with a little modification; Yamamoto was not concerned with secondary flow and used a different definition of the Nusselt number.

The problem of free convection induced by a concentrated heat source suddenly embedded in an infinite, fluid-saturated porous medium was considered by Bejan (1978). In order to describe the heat balance around the source, he used the condition

$$- k(4\pi r^2)\frac{\partial T}{\partial r} \to q \quad \text{as } r \to 0\,, \tag{5.1}$$

for which we have not found any analytical solution in term of the Rayleigh number. We have replaced (5.1) by (2.27) and, as one can see from Section 2, the analysis was completed satisfactorily.

An extensive treatment of convection in a porous medium bounded by two infinite horizontal concentric cylinders was presented by Caltagirone (1976). The two-dimensional steady solutions were obtained (1) numerically and (2) by utilizing a perturbation technique with respect to the Rayleigh number. However, probably because of the grid spacing in the first case, and the small range of Rayleigh numbers considered in the second case, Caltagirone did not mention the steady two-dimensional secondary flow reported in Section 3 in this chapter.

Other examples of natural convection in porous media, similar to those treated in this chapter, may be found in Arquis and Caltagirone (1984), Beavers and Sparrow (1971), Bories (1970), Burns and Tien (1979), Caltagirone (1980), Caltagirone *et al.* (1980), Chhuon and Caltagirone (1979), Combarnous (1970), Combarnous and Bories (1975), Elder (1967, 1968), Gheorghiţa (1961, 1969), Green (1962), Kvernvold and Tyvand (1979), Lapwood (1948), McKibbin and O'Sullivan (1980, 1981), McKibbin and Tyvand (1982), Mojtabi and Caltagirone (1978, 1979a, b, 1980), Quintard (1983), Quintard and Prouvost (1982a,b), Rudraiah and Musuoka (1982), Schubert and Strauss (1982), and Strauss (1974).

The idea of considering the thermal boundary-layer approximation for large Rayleigh numbers appears for the first time in the paper by Wooding (1963). The examples for the steady case presented in Section 4 are from Wooding (1963) and Cheng and Minkowycz (1977). Transient flow was considered by Cheng and Pop (1984). The methods used to study the thermal boundary layer are similar to those used in the theory of a viscous boundary layer. In fact, the equation obtained in the steady case is exactly Blasius' equation from classical boundary-layer theory (see for example Batchelor (1970)). It is important to note that this similarity in methods and equations is formal, because the physical significance of different terms is completely different. For example, in classical boundary-layer theory the velocity is zero at the wall; for porous media, on the impervious wall the velocity is tangential and has its maximum value there. For other examples and approximations in this problem, the reader is referred to the papers of Cheng (1978), Ingham *et al.* (1982), (1983), McNabb (1966), Na and Pop (1983) and Pop and Cheng (1983).

PART II

# Multiphase Flow

CHAPTER 4

# Macroscopic Equations

Multiphase flow in porous media occurs in many branches of engineering; an example is the exploitation of oil fields by *in situ* combustion. In such problems, increasingly complex phenomena develop during the flow: diffusion in the various phases, exchange of components between the phases, heat transfer and chemical reactions. The equations used in the mathematical formulation of the laws governing these flows must account for all such phenomena.

In order to obtain equations valid at the macroscopic (or global) level, we consider the porous medium as a mixture of continua. Then at the microscopic (or local) level each phase is a continuous body. On passing from the microscopic level to the macroscopic, the method used is to take mean values, based on the principle that each property of the mean motion will be a mathematical consequence of the properties of the motion of the phases and components. Consequently a balance equation for the entropy is established directly at the macroscopic level. This procedure assumes that macroscopic quantities such as pressure, temperature and mass per unit volume are related by the same state equations as those valid at pore level under equilibrium conditions. This expresses the physical fact that the disorder in the motion due to the complex geometry of the pores must contribute to the increase of entropy only at the macroscopic level. In this macroscopic balance equation there may also appear generalized forces and fluxes. The constitutive equations relating these forces and fluxes at the macroscopic level are quite complicated. A number of simplifying assumptions are made in Section 4 in order to obtain a system of equations representing underground combustion.

In view of the scope of this book, we neglect in the presentation of this problem all interfacial phenomena. Of course, this method may be used to treat any phenomenon of this kind (see Marle (1982) and the bibliography cited there).

## 1. Equations for the Pore Level

We consider a porous medium containing in its pores several fluid phases denoted by letters $a$, $b$, .... The index s indicates the solid part of the medium; the solid is perfectly rigid and fixed in space, in a Galilean frame of reference. Each fluid phase may contain $n$ components, denoted by $i$, $j$, ... . These components may diffuse in the phase, may be transformed to the adjacent phases and may participate to chemical reactions. No mass transfer or chemical reaction occurs inside the solid, or at its surface. However, energy transfer by conduction may occur inside the solid, and heat may be exchanged between it and adjacent fluid phases. This exchange, according to Section 3 of Chapter 1, is instantaneous. The various components of the fluid phases are subjected to external forces, but no external torque is exerted on the medium.

Proceeding with these assumptions, we have the following equations (see Appendix 1) for a continuous medium:

$$\frac{\partial}{\partial t}(\rho c_i) + \frac{\partial}{\partial x_k}(\rho c_i v_k + j_k^i) - \sum_\alpha \nu_{i\alpha} \psi_\alpha = 0\,, \tag{1.1}$$

$$\frac{\partial}{\partial t}(\rho v_k) + \frac{\partial}{\partial x_l}(\rho\, v_l v_k - t_{kl}) - \sum_i \rho\, c_i f_k^i = 0\,, \tag{1.2}$$

$$\frac{\partial}{\partial t}(\rho \varepsilon) + \frac{\partial}{\partial x_l}(\rho \varepsilon\, v_l + q_l) - t_{kl}\frac{\partial v_l}{\partial x_k} - \sum_i j_l^i f_l^i = 0\,, \tag{1.3}$$

which express the balance of mass, linear momentum and energy. In these equations $\rho$ is the mass per unit volume, $c_i$ the mass concentration of component $i$, $v_k$ the $k$th component of the velocity $\mathbf{v}$, $\mathbf{j}^i$ the diffusion mass flux of component $i$, $\varepsilon$ the internal energy, $t_{kl}$ the total stress tensor, $\mathbf{q}$ the heat flux, $\mathbf{f}^i$ the external force exerted per unit mass of component $i$, $\psi_\alpha$ the intensity of chemical reaction $\alpha$ per unit volume and $\nu_{i\alpha}$ the stoichiometric coefficients of this reaction. Also,

$$\sum_i \nu_{i\alpha} = 0\,, \tag{1.4}$$

$$\sum_i c_i = 1\,, \tag{1.5}$$

$$\sum_i \mathbf{j}^i = 0\,, \tag{1.6}$$

$$t_{kl} = t_{lk}\,. \tag{1.7}$$

The summation of (1.1) over $i$ yields the continuity equation

$$\frac{\partial \rho}{\partial t} + \operatorname{div}(\rho \mathbf{v}) = 0\,. \tag{1.8}$$

All quantities appearing in the above equations will be indexed by the letter $a$ for the fluid phase and by s for the solid. (For example, $\rho_a$ is the mass per unit volume in phase $a$.) By definition, any quantity indexed by $a$ is equal to zero outside phase $a$. Consequently, these quantities are continuous with continuous derivatives in each phase, and have jump discontinuities across interfaces. Derivatives appearing in the above equations must be taken in the sense of distributions (Schwartz (1965)).

For the solid, assumed perfectly rigid and at rest, we have only the balance of energy:

$$\frac{\partial}{\partial t}(\rho_s \varepsilon_s) + \frac{\partial q_k^s}{\partial x_k} = 0 . \tag{1.9}$$

## 2. Macroscopic Balance Equations

### 2.1. MACROSCOPIC QUANTITIES

The general motive for defining a macroscopic quantity at point $x$ and time $t$ is to take the mean value of the corresponding quantity valid at microscopic level, over some elementary volume around point $x$ and at time $t$. The best mathematical tool for this is the convolution product with a weight function.

Let $\alpha$ be a positive function with compact support in $\mathbb{R}^3$, such that

$$\int_{\mathbb{R}^3} \alpha(x)\, \mathrm{d}x = 1 . \tag{2.1}$$

Let $f(x, t)$ be a microscopic quantity, which can be a scalar, vector or tensor function of space and time.

By definition the macroscopic quanitity, associated with the microscopic quanitity $f(x, t)$ is the convolution product $f * \alpha$ with respect to the space variables:

$$(f * \alpha)(x, t) = \int_{\mathbb{R}^3} f(x - \xi, t)\, \alpha(\xi)\, \mathrm{d}\xi = \int_{\mathbb{R}^3} f(\xi, t)\, \alpha(x - \xi)\, \mathrm{d}\xi . \tag{2.2}$$

In practice, the choice of $\alpha$ must be made, for each case, in accordance with physical considerations. For example, when $\alpha$ is equal to $1/V$ in a ball of volume $V$ centred on the origin, and equal to zero outside this ball, $f * \alpha$ is the ordinary mean value of $f$, at time $t$, over a ball of volume $V$ centred on the point $x$. However, this particular choice of $\alpha$ may not be very advantageous because $f * \alpha$ may have discontinuous derivatives. It seems better to choose $\alpha$ to be of class $C^\infty$; i.e., continuous and with continuous derivatives of all orders. This ensures that $f * \alpha$ will also be of class $C^\infty$, with respect to the space variable $x$.

In order to understand this definition, it is helpful to remark that in statistical mechanics the macroscopic quantities are defined as averages over all sets of systems macroscopically identicals. A microscopic quantity $f(x)$ is, from this point of view, the realization of a particular random function, while the corresponding macroscopic quanitity is the mean of this random function. The definition of a macroscopic quantity as a spatial-average weighting with a function $\alpha(x)$ is a particular case of the previous definition. Let $\alpha$ be a determinate porous medium. We consider the family of all porous media derived from it by translation by a vector $\boldsymbol{\xi}$. With each of these translated porous media is associated a probability density $\alpha(\boldsymbol{\xi})$. If $f(x)$ is a microscopic quantity, we define a random function which has the possible realization $f(\mathbf{x} + \boldsymbol{\xi})$, with probability density $\alpha(\boldsymbol{\xi})$. The mean value of this random function is $f * \alpha$, to coincide with (2.2)

Our equations also involve the derivatives of the microscopic quantities. The differentiation formulae for the macroscopic quantities are

$$\frac{\partial}{\partial x_i}(f * \alpha) = \left( \left\{ \frac{\partial f}{\partial x_i} \right\} + [f]\, n_i\, \delta_{ab} \right) * \alpha\,, \tag{2.3}$$

$$\frac{\partial}{\partial t}(f * \alpha) = \left( \left\{ \frac{\partial f}{\partial t} \right\} - [f]\, \mathbf{n} \mathbf{V}\, \delta_{ab} \right) * \alpha\,, \tag{2.4}$$

where $\mathbf{V}$ is the velocity of the interface $S_{ab}$ of separation between phases $a$ and $b$, $\mathbf{n}$ is the unit vector normal to $S_{ab}$ directed towards $b$, $[f]$ is the jump of $f$ on the surface $S_{ab}$, $\left\{ \frac{\partial f}{\partial x_i} \right\}$ and $\left\{ \frac{\partial f}{\partial t} \right\}$ are the derivatives of $f$ in the usual sense, and $\delta_{ab}$ is the Dirac distribution in the three-dimensional space with respect to $S_{ab}$.

From the mathematical point of view, the definition of a macroscopic quantity makes sense for any microscopic quanitity, but the quantity thus obtained does not always have a physical meaning. For example, if $f$ is the density, its integral over a volume $V$ does have a physical meaning: it is precisely the mass of the fluid phase in the volume $V$. For such quantities the definition is correct. In contrast, if $f$ is the temperature, its integral over a volume $V$ has no physical meaning. Similarly, if $f$ is the velocity, its integral over a volume $V$ has no physical meaning, but $\rho v$ is the quantity of motion, and its integral does have a precise physical meaning. Therefore we can define the macroscopic quantities only for those microscopic ones whose integral over a volume $V$ does has a physical meaning.

## 2.2. MACROSCOPIC BALANCE EQUATIONS

To obtain macroscopic equations, one has only to take the convolution product with $\alpha$ of equations valid at the pore level, established in Section 1.

For any fluid phase $a$, let $\chi_a(x, t)$ be the function equal to 1 in phase $a$ and to 0 elsewhere. For the solid, $\chi_s(x, t)$, which is independent of $t$, is similarly defined.

The porosity $n$ of the medium and the saturation $S_a$ in the fluid phase $a$ are defined by

$$n = (1 - \chi_s) * \alpha\,, \tag{2.5}$$

$$nS_a = \chi_a * \alpha\,. \tag{2.6}$$

Now, consider equation (1.1) written for phase $a$ only, and take its convolution product with α. Using (2.3) and (2.4), one obtains the macroscopic balance equation for the *i*th component in phase *a* as

$$\frac{\partial}{\partial t}(nS_aR_aC_{ia}) + \frac{\partial}{\partial x_k}(nS_aR_aC_{ia}U_k^a + J_k^{ia}) + B_{ias} + \sum_{a \neq b} B_{iab} -$$

$$- \; nS_a \sum_\alpha \nu_{i\alpha} \Psi_\alpha = 0, \tag{2.7}$$

where $R_a$ is the macroscopic mass per unit volume of phase *a*,

$$nS_aR_a = \rho_a * \alpha\,, \tag{2.8}$$

$C_{ia}$ is the macroscopic mass concentration of the *i*th component in phase *a*,

$$nS_aR_a\, C_{ia} = (\rho_a C_{ia}) * \alpha\,, \tag{2.9}$$

$\mathbf{U}^a$ is the macroscopic velocity of phase *a*,

$$nS_aR_a\mathbf{U}^a = (\rho_a\mathbf{v}^a) * \alpha\,, \tag{2.10}$$

$\mathbf{J}^{ia}$ is the macrosopic diffusion-dispersion mass flux of the *i*th component in phase *a*,

$$J_k^{ia} = (j_k^{ia}) * \alpha + (\rho_a\,(c_{ia} - C_{ia})\,(v_k^a - U_k^a)) * \alpha\,, \tag{2.11}$$

and $B_{iab}$ is the mass of the *i*th component which leaves phase *a* through $S_{ab}$ per unit time and unit volume of porous medium,

$$B_{iab} = ([\rho_a\, C_{ia}\,(v_k^a - V_k) + j_k^{ia}]\, n_k\, \delta_{ab}) * \alpha\,; \tag{2.12}$$

a similar definition holds for $B_{ias}$. The term $\Psi_\alpha$ is the macroscopic rate of reaction α, per unit of volume of phase *a*, defined by

$$nS_a\Psi_\alpha = \psi_\alpha * \alpha\,. \tag{2.13}$$

It is interesting to note that the macroscopic diffusion – dispersion mass flux is not simply the mean value of the corresponding microscopic flux; it includes also effects of differences between the local convection flux $\rho_a c_{ia} v_k^a$, and its mean value $nS_aR_aC_{ia}U_k^a$, which represent the fluctuation of this flux.

The summation of (2.7) over *i* yields the macroscopic continuity equation for phase *a*:

$$\frac{\partial}{\partial t}(nS_aR_a) + \frac{\partial}{\partial x_k}(nS_aR_aU_k^a) + B_{as} + \sum_{a \neq b} B_{ab} = 0. \tag{2.14}$$

It is clear that we have used the relations

$$\sum_i C_{ia} = 1, \qquad \sum_i \mathbf{J}^{ia} = 0, \qquad B_{ab} = \sum_i B_{iab}, \qquad B_{as} = \sum_i B_{ias}\,.$$

The same procedure applied to Equations (1.2) and (1.3) leads to the following equations:

$$\frac{\partial}{\partial t}(nS_aR_aU_k^a) + \frac{\partial}{\partial x_l}(nS_aR_aU_k^aU_l^a - T_{kl}^a) +$$

$$+ \sum_{a\neq b}(B_{ab}U_k^a + \mathfrak{F}_k^{ab}) + B_{as}U_k^a + \mathfrak{F}_k^{as} - nS_aR_a\sum_i C_{ia}f_k^i = 0\,, \tag{2.15}$$

$$\frac{\partial}{\partial t}(nS_aR_aE_a) + \frac{\partial}{\partial x_l}(nS_aR_aE_aU_l^a + Q_l^a) - T_{kl}^a\frac{\partial U_l^a}{\partial x_k} +$$

$$+ \sum_{a\neq b}\left(\left(E_a + \frac{P_a}{R_a}\right)B_{ab} + D_{ab} - \mathfrak{F}_k^{ab}U_k^a\right) +$$

$$+ \left(E_a + \frac{P_a}{R_a}\right)B_{as} + D_{as} + P_a\frac{\partial(nS_a)}{\partial t} - \mathfrak{F}_k^{as}U_k^a - \sum_i J_k^{ia}f_k^i = 0\,. \tag{2.16}$$

The macroscopic stress tensor $T_{kl}^a$ is defined by

$$T_{kl}^a = (t_{kl}^a) * \alpha + (\rho_a(v_k^a - U_k^a)(v_l^a - U_l^a)) * \alpha\,. \tag{2.17}$$

In the same way, we have the definitions

$$B_{ab}U_k^a + \mathfrak{F}_k^{ab} = ([t_{kl}^a - \rho_a v_k^a(v_l^a - V_l)]\, n_l\,\delta_{ab}) * \alpha\,, \tag{2.18}$$

$$B_{as}U_k^a + \mathfrak{F}_k^{as} = ([t_{kl}^a - \rho_a v_k^a v_l^a]\, n_l\,\delta_{ab}) * \alpha\,, \tag{2.19}$$

$$nS_aR_aE_a = (\rho_a\,\varepsilon_a) * \alpha\,, \tag{2.20}$$

$$Q_l^a = (q_l^a) * \alpha + (\rho_a(\varepsilon_a - E_a)\,(v_l^a - U_l^a)) * \alpha - (v_k^a(t_{kl}^a - T_{kl}^a)) * \alpha\,, \tag{2.21}$$

$$\left(E_a + \frac{P_a}{R_a}\right)B_{ab} + D_{ab} + P_a\,W_{ab} = \left(\left[\rho_a\left(\varepsilon_a + \frac{P_a}{\rho_a}\,\delta_{kl}\right)(v_l^a - V_l) +\right.\right.$$

$$\left.\left. + \, q_l^a - v_k^a\,\tau_{kl}^a + P_a\,\delta_{kl}\,V_l\right] n_l\,\delta_{ab}\right) * \alpha\,, \tag{2.22}$$

$$W_{ab} = ((V_l\,n_l)\,\delta_{ab}) * \alpha\,, \tag{2.23}$$

$$\sum_{a\neq b} W_{ab} = \frac{\partial(nS_a)}{\partial t}\,, \tag{2.24}$$

$$\left(\sum_i \rho_a c_{ia} f_k^i\right) * \alpha = \sum_i((\rho_a\,c_{ia}) * \alpha)\,f_k^i = nS_aR_a\sum_i C_{ia}\,f_k^i\,, \tag{2.25}$$

$$\left(\sum_i(\rho_a c_{ia} v_l^a + j_l^{ia})\,f_l^i\right) * \alpha = \sum_i(nS_aR_aC_{ia}U_l^a + J_l^{ia})\,f_l^i\,. \tag{2.26}$$

The remark made above also applies to these macroscopic quantities: they are not simply the mean values of the corresponding microscopic quantities. For instance, (2.17) or (2.21) incorporates the fluctuations of the velocity or mass per unit volume. Rather more complicated is the definition of the quantity of heat leaving phase $a$ through $S_{ab}$ per unit time and unit volume of porous medium, (2.22). The macroscopic thermodynamic quantities $R_a$, $E_a$ and $C_{ia}$ have been defined, so it is legitimate to assume that all other macroscopic quantities are related to $R_a$, $E_a$ and $C_{ia}$ by the same constitutive equations that are valid at the microscopic level under equilibrium conditions. We then have a macroscopic pressure $P_a$, a macroscopic temperature $\theta_a$, and (2.22).

The macroscopic stress tensor is defined by

$$T^a_{kl} = -\, nS_aP_a\delta_{kl} + \tau^a_{kl}\,. \tag{2.27}$$

Equation (2.23) gives the volume swept by $S_{ab}$ per unit time and unit volume of porous medium. Using (2.4) and (2.6), we have (2.24). Similar definitions hold for $D_{as}$ and $W_{as}$.

It has been also assumed that the external body force $\mathbf{f}^i$ is uniform through all phases (like the gravity), and consequently we have (2.25) and (2.26).

In the same way, the convolution product with $\alpha$ of equation (1.9) leads to the macroscopic energy balance equation for the solid:

$$\frac{\partial}{\partial t}((1-n)R_sE_s) + \frac{\partial Q^s_l}{\partial x_l} + \sum_a D_{sa} = 0\,, \tag{2.28}$$

where

$$(1-n)\,R_s = \rho_s * \alpha\,, \qquad (1-n)\,R_sE_s = (\rho_s\varepsilon_s) * \alpha\,, \qquad Q^s_l = q^s_l * \alpha\,,$$

$$D_{sa} = ([q^s_l]\, n_l\, \delta_{ab}) * \alpha\,.$$

## 3. The Macroscopic Entropy Balance Equation

We assume that all the relations between thermodynamic quantities which are valid for a medium at rest under equilibrium thermodynamic conditions remain valid at each point for the macroscopic thermodynamic quantities of the various fluid phases and of the solid. Under this hypothesis, it is possible to write for fluid phase $a$ a relation between differentials:

$$dE_a = -\, P_a\,\mathrm{d}\left(\frac{1}{R_a}\right) + \theta_a\,\mathrm{d}\eta_a + \sum_i \mu_{ia}\,\mathrm{d}C_{ia}\,, \tag{3.1}$$

where $\theta_a$ is the macroscopic temperature, $\eta_a$ is the specific entropy, and $\mu_{ia}$ is the macroscopic thermodynamic potential of the $i$th component in phase $a$. Of course, we also have the Gibbs – Duhem equation

$$E_a + \frac{P_a}{R_a} - \theta_a\eta_a - \sum_i \mu_{ia}C_{ia} = 0\,. \tag{3.2}$$

From the two preceding equations we have

$$\mathrm{d}(R_a E_a) = \theta_a \,\mathrm{d}(R_a \eta_a) + \sum_i \mu_{ia}\,\mathrm{d}(R_a C_{ia})\,. \tag{3.3}$$

In the solid phase the corresponding equation is simply

$$\mathrm{d}(R_s E_s) = \theta_s \,\mathrm{d}(R_s \eta_s)\,. \tag{3.4}$$

In order to obtain the final form of the macroscopic entropy balance equation, we begin with the terms

$$\frac{\partial}{\partial t}(nS_a R_a E_a) + \frac{\partial}{\partial x_k}(nS_a R_a E_a U_k^a)\,,$$

using the equations (2.7), (2.14), (3.1) and (3.2); introducing it into the equation (2.16) we derive

$$\begin{aligned}
&\frac{\partial}{\partial t}(nS_a R_a \eta_a) + \frac{\partial}{\partial x_k}\left(nS_a R_a \eta_a\, U_k^a - \sum_i \frac{\mu_{ia}}{\theta_a} J_k^{ia} + \frac{Q_k^a}{\theta_a}\right) - \frac{\tau_{kl}^a}{\theta_a}\frac{\partial U_k^a}{\partial x_l} - \\
&- Q_k^a \frac{\partial}{\partial x_k}\left(\frac{1}{\theta_a}\right) + \sum_i J_k^{ia}\left[\frac{\partial}{\partial x_k}\left(\frac{\mu_{ia}}{\theta_a}\right) - \frac{f_k^i}{\theta_a}\right] + \\
&+ \frac{nS_a}{\theta_a}\sum_\alpha \psi_\alpha \left(\sum_i \mu_{ia} v_{ia}\right) + \eta_a B_{ab} - \sum_i \frac{\mu_{ia}}{\theta_a}(B_{ias} - C_{ia} B_{as}) + \\
&+ \frac{D_{as}}{\theta_a} + \sum_{a \neq b}\left[\eta_a B_{ab} - \sum_i \frac{\mu_{ia}}{\theta_a}(B_{iab} - C_{ia} B_{ab}) + \frac{D_{ab}}{\theta_a}\right] - \\
&- \frac{U_k^a}{\theta_a}\left[\mathfrak{F}_k^{as} + \sum_{a \neq b} \mathfrak{F}_k^{ab} + P_a \frac{\partial (nS_a)}{\partial x_k}\right] = 0\,.
\end{aligned} \tag{3.5}$$

Similarly, for the solid phase we have

$$\frac{\partial}{\partial t}[(1-n)\, R_s \eta_s] + \frac{\partial}{\partial x_k}\left(\frac{Q_k^s}{\theta_s}\right) - Q_k^s \frac{\partial}{\partial x_k}\left(\frac{1}{\theta_s}\right) + \frac{1}{\theta_s}\sum_a D_{as} = 0\,. \tag{3.6}$$

The total macroscopic entropy balance equation may be obtained by summing the macroscopic entropy balance equations for all fluid and solid phases. But this equation is more complicated, and it seems that in this form it is difficult to manipulate. For this reason we discuss in Section 4 some particular cases and simplyfing assumptions.

It is clear that this equation has the usual form for an entropy balance equation: a time derivative of entropy density plus the divergence of an entropy flux equals entropy source density.

## 4. Simplifying Assumptions

First we consider the system of equations for just one fluid phase, f. We have $S = 1$. We denote by $\rho$ the macroscopic mass per unit volume, and we preserve the index s for the solid phase. Equations (2.7), (2.14), (2.15), (2.16), (2.28), (3.5) and (3.6) then take the form

$$\frac{\partial}{\partial t}(n\rho C_i) + \frac{\partial}{\partial x_k}(n\rho C_i U_k + J_k^i) - n\sum_\alpha \nu_{i\alpha}\,\psi_\alpha = 0\,, \tag{4.1}$$

$$\frac{\partial}{\partial t}(n\rho) + \frac{\partial}{\partial x_k}(n\rho U_k) = 0\,, \tag{4.2}$$

$$\frac{\partial}{\partial t}(n\rho U_k) + \frac{\partial}{\partial x_l}(n\rho U_k U_l - T_{kl}) - \mathfrak{F}_k^{\mathrm{fs}} - n\rho\sum_i C_i f_k^i = 0\,, \tag{4.3}$$

$$\frac{\partial}{\partial t}(n\rho E) + \frac{\partial}{\partial x_l}(n\rho E U_l + Q_l^{\mathrm{f}}) - T_{kl}\frac{\partial U_l}{\partial x_k} + P\frac{\partial n}{\partial t} + D_{\mathrm{fs}} - \mathfrak{F}_k^{\mathrm{fs}}\,U_k -$$

$$-\sum_i J_k^i f_k^i = 0\,, \tag{4.4}$$

$$\frac{\partial}{\partial t}(\,(1-n)\,R_{\mathrm{s}}E_{\mathrm{s}}) + \frac{\partial Q_l^{\mathrm{s}}}{\partial x_l} + D_{\mathrm{sf}} = 0, \tag{4.5}$$

$$D_{\mathrm{fs}} + D_{\mathrm{sf}} = 0\,, \tag{4.6}$$

$$\frac{\partial}{\partial t}(n\rho\eta) + \frac{\partial}{\partial x_l}\left(n\rho\eta U_k - \sum_i \frac{\mu_i}{\theta_{\mathrm{f}}}J_k^i + \frac{Q_k^{\mathrm{f}}}{\theta_{\mathrm{f}}}\right) - \frac{\tau_{kl}}{\theta_{\mathrm{f}}}\frac{\partial U_k}{\partial x_l} -$$

$$-\;Q_k^{\mathrm{f}}\frac{\partial}{\partial x_k}\left(\frac{1}{\theta_{\mathrm{f}}}\right) + \sum_i J_k^i\left[\frac{\partial}{\partial x_k}\left(\frac{\mu_i}{\theta_{\mathrm{f}}}\right) - \frac{f_k^i}{\theta_{\mathrm{f}}}\right] + \frac{n}{\theta_{\mathrm{f}}}\sum_\alpha \psi_\alpha\left(\sum_i \mu_i\nu_{i\alpha}\right) +$$

$$+\;\frac{D_{\mathrm{fs}}}{\theta_{\mathrm{f}}} - \frac{U_k}{\theta_{\mathrm{f}}}\left[\mathfrak{F}_k^{\mathrm{fs}} + P\frac{\partial n}{\partial x_k}\right] = 0, \tag{4.7}$$

$$\frac{\partial}{\partial t}[(1-n)R_{\mathrm{s}}\eta_{\mathrm{s}}] + \frac{\partial}{\partial x_k}\left(\frac{Q_k^{\mathrm{s}}}{\theta_{\mathrm{s}}}\right) - Q_k^{\mathrm{s}}\frac{\partial}{\partial x_k}\left(\frac{1}{\theta_{\mathrm{s}}}\right) + \frac{D_{\mathrm{sf}}}{\theta_{\mathrm{s}}} = 0\,. \tag{4.8}$$

The sum of equations (4.7) and (4.8) give us the balance equation for the entropy of the system in the form

$$\frac{\partial}{\partial t}[n\rho\eta + (1-n)R_{\mathrm{s}}\eta_{\mathrm{s}}] + \frac{\partial}{\partial x_k}\left(n\rho\eta U_k + \frac{Q_k^{\mathrm{f}}}{\theta_{\mathrm{f}}} + \frac{Q_k^{\mathrm{s}}}{\theta_{\mathrm{s}}} - \sum_i\frac{\mu_i}{\theta_{\mathrm{f}}}J_k^i\right) -$$

$$- Q_k^{\mathrm{f}} \frac{\partial}{\partial x_k}\left(\frac{1}{\theta_{\mathrm{f}}}\right) - Q_k^{\mathrm{s}} \frac{\partial}{\partial x_k}\left(\frac{1}{\theta_{\mathrm{s}}}\right) +$$

$$+ D_{\mathrm{fs}}\left(\frac{1}{\theta_{\mathrm{f}}} - \frac{1}{\theta_{\mathrm{s}}}\right) - \frac{\tau_{kl}}{\theta_{\mathrm{f}}}\frac{\partial U_k}{\partial x_l} + \frac{n}{\theta_{\mathrm{f}}}\sum_{\alpha} \psi_\alpha \left(\sum_i \mu_i \nu_{i\alpha}\right) +$$

$$+ \sum_i J_k^i \left[\frac{\partial}{\partial x_k}\left(\frac{\mu_i}{\theta_{\mathrm{f}}}\right) - \frac{f_k^i}{\theta_{\mathrm{f}}}\right] - \frac{U_k}{\theta_{\mathrm{f}}}\left[\mathfrak{F}_k^{\mathrm{fs}} + P\frac{\partial n}{\partial x_k}\right] = 0\,. \tag{4.9}$$

With the entropy balance equation in the form (4.9), it is easy to observe that the entropy source density $\sigma$ per unit volume of porous medium is

$$\sigma = Q_k^{\mathrm{f}} \frac{\partial}{\partial x_k}\left(\frac{1}{\theta_{\mathrm{f}}}\right) + Q_k^{\mathrm{s}} \frac{\partial}{\partial x_k}\left(\frac{1}{\theta_{\mathrm{s}}}\right) + D_{\mathrm{fs}}\left(\frac{1}{\theta_{\mathrm{s}}} - \frac{1}{\theta_{\mathrm{f}}}\right) +$$

$$+ \frac{\tau_{kl}}{\theta_{\mathrm{f}}}\frac{\partial U_k}{\partial x_l} - \frac{n}{\theta_{\mathrm{f}}}\sum_{\alpha} \psi_\alpha \left(\sum_i \mu_i \nu_{i\alpha}\right) -$$

$$- \sum_i J_k^i \left[\frac{\partial}{\partial x_k}\left(\frac{\mu_i}{\theta_{\mathrm{f}}}\right) - \frac{f_k^i}{\theta_{\mathrm{f}}}\right] + \frac{U_k}{\theta_{\mathrm{f}}}\left(\mathfrak{F}_k^{\mathrm{fs}} + P\frac{\partial n}{\partial x_k}\right). \tag{4.10}$$

Of course, the second law of thermodynamics may be expressed as

$$\sigma \geqslant 0. \tag{4.11}$$

The standard way to obtain the constitutive equations is then to express the heat fluxes in the fluid and in the solid, the viscous stress tensor, the quantity of heat leaving the fluid phase through the solid phase, the rate of reaction $\alpha$ and the effective force of interaction between the fluid and the solid, all as functions of the independent variables (the temperatures, the relative velocity, the difference from chemical equilibrium for the $\alpha$ reaction and their gradients). Imposing the condition (4.11), we obtain thermodynamic restrictions on the form of these constitutive equations.

In order to simplify the computation, for practical purposes it is sufficient to observe that the form of $\sigma$ given by (4.10) is a sum of products of different fluxes and independent variables. By taking the constitutive equations in linear form and with coefficients such that the total entropy source $\sigma$ has a positive quadratic form, we obtain the desired result.

To illustrate this, we consider single fluid phase of only one component. No chemical reactions occur. The temperature is not assumed to be uniform, but the difference between the macroscopic temperatures for the solid and the fluid is assumed to be negligible at each point of the porous medium. This means that heat is exchanged between the fluid and the solid sufficiently rapidly to equalize the local macroscopic temperatures (see also the hypothesis from Chapter 1). It is therefore not necessary to distinguish between the heat flux in the fluid and in the solid. We then define the total macroscopic heat flux as the sum of heat fluxes in the fluid and in the solid, and Equation (4.10) becomes

$$\sigma = \frac{\tau_{kl}}{\theta}\frac{\partial U_k}{\partial x_l} + Q_k \frac{\partial}{\partial x_k}\left(\frac{1}{\theta}\right) + \frac{U_k}{\theta}\left(\mathfrak{F}_k^{\mathrm{fs}} + P\frac{\partial n}{\partial x_k}\right) . \tag{4.12}$$

If we neglect cross-effects, then Equation (4.12) suggests the following form for the constitutive equations:

$$U_k = \Lambda\left(\mathfrak{F}_k^{\mathrm{fs}} + P\frac{\partial n}{\partial x_k}\right) , \tag{4.13}$$

$$Q_k = -\lambda\frac{\partial \theta}{\partial x_k} , \tag{4.14}$$

$$\tau_{kl} = L\frac{\partial U_i}{\partial x_i}\delta_{kl} + \frac{1}{2}M\left(\frac{\partial U_k}{\partial x_l} + \frac{\partial U_l}{\partial x_k}\right) , \tag{4.15}$$

where $\Lambda$, $\lambda$, $L$ and $M$ are phenomenological coefficients. Note that (4.15) is valid only under the hypothesis that the porous medium is macroscopically isotropic.

In this case the other equations of the problem take the form

$$\frac{\partial}{\partial t}(n\rho) + \frac{\partial}{\partial x_k}(n\rho U_k) = 0 , \tag{4.16}$$

$$\frac{\partial}{\partial t}(n\rho U_k) + \frac{\partial}{\partial x_l}(n\rho U_k U_l - T_{kl}) - \mathfrak{F}_k^{\mathrm{fs}} - n\rho f_k = 0 , \tag{4.17}$$

$$\frac{\partial}{\partial t}(n\rho E + (1-n) R_s E_s) + \frac{\partial}{\partial x_l}(n\rho E U_l + Q_l) + P\frac{\partial n}{\partial t} - T_{kl}\frac{\partial U_l}{\partial x_k} - $$

$$- \mathfrak{F}_k^{\mathrm{fs}} U_k = 0 . \tag{4.18}$$

Elimination of $\mathfrak{F}_k^{\mathrm{fs}}$, $P\frac{\partial n}{\partial x_k}$ and $\tau_{kl}$ between these equations leads to a generalized form of Darcy's law.

For the usual form of Darcy's law it is also necessary to neglect the viscous stress tensor and the inertia term. Thus, from (4.17) we have

$$U_k = n\Lambda\left[\frac{\partial P}{\partial x_k} - \rho f_k + \rho\frac{\partial U_k}{\partial t}\right] \tag{4.19}$$

which is the classicial form of Darcy's law. This equation shows that the phenomenological coefficient $\Lambda$ is related to the permeability $k$, the porosity $n$ of the medium and the viscosity $\mu$ of the fluid by the equation

$$\Lambda = -\frac{k}{\mu n^2} . \tag{4.20}$$

For the energy equation, it is sufficient to observe that

$$P\frac{\partial n}{\partial t} - T_{kl}\frac{\partial U_l}{\partial x_k} + \mathfrak{F}_k^{\text{fs}}\, U_k = -\frac{nP}{\rho}\frac{\partial \rho}{\partial t} - \frac{nP}{\rho}U_k\frac{\partial \rho}{\partial x_k} - PU_k\frac{\partial n}{\partial x_k}\ ,$$

and then we have

$$\rho\,\frac{\mathrm{d}E}{\mathrm{d}t} + \frac{1}{n}\frac{\partial}{\partial t}((1-n)\,R_s\,E_s) - \frac{1}{n}\frac{\partial}{\partial x_k}\left(\lambda\,\frac{\partial \theta}{\partial x_k}\right] -$$

$$-\frac{nP}{\rho}\frac{\partial \rho}{\partial t} - \frac{P}{\rho}U_k\frac{\partial \rho}{\partial x_k} - \mu k^{-1}\,U_k\,U_k = 0\,, \tag{4.21}$$

which agrees with Equation (3.24) in Chapter 1. It is clear that this equation contains a convection term, a conduction term, a compression term and a viscous-dissipation term. The phenomenological coefficients $\Lambda$ and $\lambda$ are in fact tensors.

## 5. Comments and Bibliographical Notes

The equations governing multiphase flow in porous media are usually written without any rigorous motivation. They are generally obtained by analogy with those governing the same phenomena in a continuous medium.

The present derivation of the macroscopic equations follows the procedure used by Ene (1981) and Marle (1980, 1982). Other, more rigorous theoretical foundations for these macroscopic equations are provided, taking as an example the simplest case of single-phase flow, by Matheron (1967) or, taking multiphase flow, by Slattery (1967, 1969) and Whitaker (1967, 1969).

For other results in the field we refer the reader to Bear (1972), Nigmatulin (1979) and Williams (1978).

CHAPTER 5

# Underground Combustion

## 1. Different Models

The general equations obtained in the previous chapter may be simplified to give the equations for underground combustion. For this we consider only four phases: gas, oil, water and coke. The total number of components is $n$, and they will be present in any fluid phase. Note that coke is a solid phase.

The more complicated problem is that concerning the pressure. If a porous medium contains various imiscible phases, the pressure is different in each of these phases. However, if we accept this we are obliged to consider different thermodynamic potentials, and consequently to introduce in the energy equation new terms connected with interface phenomena. The equations will then be very complicated and difficult to manipulate. For this reason, we assume a unique pressure $P$ in our computation of thermodynamic properties of the phases. This pressure is used in the energy-balance equation and in the constitutive equations which permit us to calculate the mass, viscosity and thermodynamic potential of each phase. However, for the pressure gradient we allow each phase to have its own pressure, and in the constitutive equation which gives the motion of a phase and in the balance of momentum equation, we introduce a term $P_a$.

As a consequence of this assumption the thermodynamic potentials of a component are the same in any phase where this component is present. We do not make a difference between chemical reactions in a phase or at the interface between phases.

We neglect the interfacial energies. We assume that there is only one temperature for all phases, and also for the solid part of the medium (see also the remarks made in Section 3 of Chapter 1).

The macroscopic stress tensor contains only the phase pressure, so the viscous stress tensors are negligibly small. All external volume forces $\mathbf{f}^i$ are equal to the gravity $\mathbf{g}$.

Hence the macroscopic mass-balance equation ((2.7) in Chapter 4) is now written for all phases in the form

$$\frac{\partial}{\partial t}\sum_a (nS_aR_aC_{ia}) + \frac{\partial}{\partial x_k}\sum_a (nS_aR_aC_{ia}U_k^a + J_k^{ia}) - \sum_\alpha \nu_{i\alpha}\psi_\alpha = 0\,. \tag{1.1}$$

Taking into account Equation (2.14) in Chapter 4, and the expression

$T^a_{kl} = -nS_aP_a\delta_{kl}$. Equation (2.15) in Chapter 4 gives, in vectorial form,

$$nS_aR_aD_a\mathbf{U}^a + \text{grad}\,(nS_aP_a) + \left(\sum_{a\neq b} \mathfrak{F}^{ab} + \mathfrak{F}^{as} - nS_aR_a\mathbf{g}\right) = 0\,, \tag{1.2}$$

where

$$D_a = \frac{\partial}{\partial t} + \mathbf{U}^a\,\text{grad}.$$

According to the previous hypotheses we may write only one equation for the balance of energy. For that we take the sum over all phases of Equations (2.16) and (2.28) in Chapter 4. The terms containing $B_{ab}$, $B_{as}$, $D_{ab}$ and $D_{as}$ vanish if we take the sum over all phases, and also, because the external force is equal to $\mathbf{g}$, we have $\Sigma_i \mathbf{J}^{ia} = 0$. Then, introducing the specific enthalpy.

$$h_a = E_a + \frac{P_a}{R_a}\,, \tag{1.3}$$

we have

$$\frac{\partial}{\partial t}\left[\sum_a nS_aR_aE_a + (1-n)R_sE_s\right] + \frac{\partial}{\partial x_l}\left[\sum_a nS_aR_ah_aU^a_l + Q_l\right] +$$

$$+ \sum_a P_a\frac{\partial(nS_a)}{\partial t} - \sum_a U^a_k\left[\sum_{a\neq b} \mathfrak{F}^{ab}_k + \mathfrak{F}^{as}_k + \frac{\partial}{\partial x_k}(nS_aP_a)\right] = 0\,, \tag{1.4}$$

where the total heat flux is represented by

$$Q_l = \sum_a Q^a_l + Q^s_l\,. \tag{1.5}$$

In Equation (1.4) the term $\sum_a P_a \dfrac{\partial(nS_a)}{\partial t}$ is equal to zero, because we use only one pressure in this equation, and $\Sigma_a S_a = 1$. Then, taking into account Equation (1.2), the final form of the energy-balance equation, in vectorial form, is

$$\frac{\partial}{\partial t}\left[\sum_a nS_aR_aE_a + (1-n)R_sE_s\right] + \text{div}\left[\sum_a nS_aR_ah_a\mathbf{U}^a + \mathbf{Q}\right] +$$

$$+ \sum_a nS_aR_a\mathbf{U}^a[D_a\mathbf{U}^a - \mathbf{g}] = 0. \tag{1.6}$$

The total macroscopic entropy-balance equation is obtained by summing up the macroscopic entropy-balance equations for all fluid phases ((3.5) in Chapter 4) and for the solid phase ((3.6) in Chapter 4). Using the assumptions set out at the beginning of this section, we obtain

$$\frac{\partial}{\partial t}\left[\sum_a nS_aR_a\eta_a + (1-n)R_s\eta_s\right] +$$

$$+ \operatorname{div} \left[ \frac{\mathbf{Q}}{\theta} + \sum_a \left( nS_a R_a \eta_a \mathbf{U}^a - \sum_i \frac{\mu_i}{\theta} \mathbf{J}^{ia} \right) - \right.$$

$$\left. - \mathbf{Q} \operatorname{grad} \left( \frac{1}{\theta} \right) \right] + \sum_a \left[ \sum_i \mathbf{J}^{ia} \left( \operatorname{grad} \left( \frac{\mu_i}{\theta} \right) - \frac{1}{\theta} \mathbf{g} \right) \right] -$$

$$- \sum_a \frac{1}{\theta} \mathbf{U}^a \left[ \sum_{a \neq b} \mathfrak{F}^{ab} + \mathfrak{F}^{as} + P_a \operatorname{grad} (nS_a) \right] + \frac{1}{\theta} \sum_\alpha \psi_\alpha \left( \sum_i \mu_i \nu_{i\alpha} \right) = 0. \tag{1.7}$$

In this form, it is easy to see that the entropy source density $\sigma$ per unit volume of porous medium is

$$\sigma = \mathbf{Q} \operatorname{grad} \left( \frac{1}{\theta} \right) + \frac{1}{\theta} \sum_a \left[ \sum_i \mathbf{J}^{ia} \left( \mathbf{g} - \theta \operatorname{grad} \left( \frac{\mu_i}{\theta} \right) \right) \right] +$$

$$+ \frac{1}{\theta} \sum_a \mathbf{U}^a \left[ \sum_{a \neq b} \mathfrak{F}^{ab} + \mathfrak{F}^{as} + P_a \operatorname{grad} (nS_a) \right] - \frac{1}{\theta} \sum_\alpha \psi_\alpha \left( \sum_i \mu_i \nu_{i\alpha} \right). \tag{1.8}$$

Now, according to the second law of thermodynamics, $\sigma \geqslant 0$. If, in Section 4 of Chapter 4, we neglect cross-effects, Equation (1.8) suggests the following constitutive equations:

$$\mathbf{Q} = - \lambda \operatorname{grad} \theta, \tag{1.9}$$

$$\mathbf{J}^{ia} = A \left[ \mathbf{g} - \theta \operatorname{grad} \left( \frac{\mu_i}{\theta} \right) \right], \tag{1.10}$$

$$\mathbf{U}^a = \frac{k_a}{n \mu_a S_a} \left[ \sum_{a \neq b} \mathfrak{F}^{ab} + \mathfrak{F}^{as} + P_a \operatorname{grad} (nS_a) \right], \tag{1.11}$$

$$\psi_\alpha = - B_\alpha \left( \sum_i \mu_i \nu_{i\alpha} \right). \tag{1.12}$$

In these equations $\lambda$ is the thermal conductivity tensor, $k_a/\mu_a$ is the mobility of the phase $a$, $A$ is a tensor which describes diffusion-dispersion effects and $B_\alpha$ is a coefficient respresenting the effects of chemical reactions. The mobility has been introduced into Equation (1.10) so that the relative permeability tensor can be used.

Elimination of $\sum_{a \neq b} \mathfrak{F}^{ab} + \mathfrak{F}^{as}$ between (1.11) and (1.2) leads to a generalized Darcy equation:

$$\mathbf{U}^a = \frac{k_a}{\mu_a} [(\operatorname{grad} P_a - R_a \mathbf{g}) + R_a D_a \mathbf{U}^a]. \tag{1.13}$$

The classical equation for the relative permeability is obtained by neglecting the term $D_a \mathbf{U}^a$.

The system of equations for underground combustion thus obtained is the following:

$$\frac{\partial}{\partial t}\left(n\sum_{\alpha} S_a R_a C_{ia}\right) + \operatorname{div}\left(\sum_{\alpha}(nS_a R_a C_{ia}\mathbf{U}^a + \mathbf{J}^{ia})\right) - \sum_{\alpha} \nu_{i\alpha}\psi_\alpha = 0\,, \tag{1.14}$$

$$\mathbf{U}^a = -\frac{k_a}{\mu_a}(\operatorname{grad} P_a - R_a\mathbf{g})\,, \tag{1.15}$$

$$\frac{\partial}{\partial t}\left[\sum_a nS_a R_a E_a + (1-n)R_s E_s\right] + \operatorname{div}\left[\sum_a nS_a R_a h_a \mathbf{U}^a + \mathbf{Q}\right] -$$

$$- n\mathbf{g}\sum_a S_a R_a \mathbf{U}^a = 0, \tag{1.16}$$

$$\mathbf{Q} = -\lambda \operatorname{grad}\theta, \tag{1.17}$$

$$\mathbf{J}^{ia} = A\left[\mathbf{g} - \theta \operatorname{grad}\left(\frac{\mu_i}{\theta}\right)\right], \tag{1.18}$$

$$\psi_\alpha = -B_\alpha\left(\sum_i \mu_i \nu_{i\alpha}\right). \tag{1.19}$$

## 2. Equations for Combustion with Reaction Zone

The equations that have to be used in this case are Equations (1.14)−(1.19) above. Equation (1.14) is the continuity equation for the constituent *i*. In practice *n* such equations, one for each constituent, have to be written. In underground combustion the number of constituents is at least six (oxygen, water, neutral gas, light and heavy hydrocarbons, and coke). By neutral gas we mean a non-condensing gas mixture which includes the nitrogen from the air, the combustion products CO and $CO_2$, and the non-condensing hydrocarbons present in the oil field before exploitation.

Equation (1.15) represents a generalized Darcy law for multiphase flow, and combines the balance of momentum and the constitutive law imposed by the second law of thermodynamics. There are at least three of these equations, one for each fluid phase (water, gas and oil). Equation (1.16) is the one energy equation to be used.

Equations (1.17)−(1.19) are the constitutive laws for the thermal conduction, the diffusion dispersion of the *i*th constituent and the kinetics of reaction α. We need three equations of the type (1.17), one equation of the type (1.18) for each constituent present in each phase, and as many equations of the type (1.19) as there are chemical reactions. In the simplest case two chemical reactions have to be taken into account: one is the combustion of the light hydrocarbon (by its reaction with the oxygen, water and non-condensing gas being formed) and the other represents in fact the coke of heavy hydrocarbons (light hydrocarbons, coke and non-condensing gas being formed) followed by the combustion of coke (gas and water being formed). Clearly, we need a mass balance for each of these reactions: $\Sigma_i \nu_{i\alpha} = 0$, in which by convention $\nu_{i\alpha}$ is positive if the $i$th constituent is formed by the reaction, and negative if it is consumed. The intensity of the chemical reaction is given by $\psi_\alpha$, the quantity $\nu_{i\alpha}\psi_\alpha$ being the mass of the $i$th constituent formed by the reaction per unit volume of porous medium and unit time.

In practice one usually neglects transport by diffusion dispersion. This implies that $\mathbf{J}^{ia} = 0$, and Equation (1.18) vanishes. If not, Equation (1.18) is usually replaced by the equation that gives $\mathbf{J}^{ia}$ in terms of the gradient of concentration. Similarly, Equation (1.19) can also be replaced. Using these different forms of the intensity of the chemical reactions does not essentially change the problem, and it may be possible that such formulations, which take into account a state far from chemical equilibrium, are more suited.

It is necessary to add some more equations to this system. Of these, the equations

$$\sum_a S_a = 1, \quad \sum_i C_{ia} = 1, \quad \sum_i \mathbf{J}^{ia} = 0$$

have already been used to deduce the complete system. (Note that Equations (1.18) are not all independent.) To these we have to add the equations of state that express the properties of the phases as functions of the independent thermodynamic variables, and the equations that express the thermodynamic equilibrium of the phases. To complete the system we need equations for the pressure differences between phases in terms of pressure, temperature and composition of phases.

This system of equations is complete from the mathematical point of view, and satisfactory from the physical point of view. Nevertheless, its solution raises very difficult computational problems, mainly because of the great number of equations and unknowns. For instance, in the one-dimensional problem, if four phases, six constituents and two chemical reactions are considered, the system will contain 28 unknowns. If diffusion-dispersion transport is neglected, the number of equations and unknowns reduces to 13, which is still a difficult problem to deal with.

### 3. The Thin Reaction Zone

The model we discuss in this section is due to Marle (1980). The main idea, similar to one from aerodynamics, is the assimilation of the reaction zone with a surface – a combustion front. If chemical reactions take place only on this surface, there will be an upstream and a downstream zone (zones I and II), with mutliphase flow and exchanges between the phases, but without chemical reactions.

This approximation is justifiable because the reaction zone is small (of the order of few metres) in comparison with the dimensions of the oil field, and is desirable because of the difficulties that arise in the numerical solution of the complete problem. These difficulties result from the necessity of considering different discretization steps in the reaction zone and in the neighbouring zones. It is clear that the number of unknowns in the problem is not reduced if the combustion zone is approximated by a surface, but it is sufficient to use a discretization with steps of equal dimensions in all zones moving in time with the combustion front.

Hence, if we suppose that the combustion front is represented by a surface which separates the zone I of the injection wells from zone II of the extraction wells, it follows that in these zones equations (1.14) – (1.18) are valid after simplifications by taking into account the absence of chemical reactions and by neglecting the effects of diffusion dispersion:

$$\frac{\partial}{\partial t}\left(n\sum_a S_a R_a C_{ia}\right) + \operatorname{div}\left(n\sum_a S_a R_a C_{ia} \mathbf{U}^a\right) = 0\,, \tag{3.1}$$

$$\mathbf{U}^a = -\frac{k_a}{\mu_a}(\operatorname{grad} P_a - R_a\,\mathbf{g})\,, \tag{3.2}$$

$$\frac{\partial}{\partial t}\left(\sum_a nS_aR_aE_a + (1-n)\,R_sE_s\right) + \operatorname{div}\left(\sum_a nS_aR_ah_a\,\mathbf{U}^a + \mathbf{Q}\right) -$$

$$- n\mathbf{g}\sum_a S_aR_a\mathbf{U}^a = 0\,, \tag{3.3}$$

$$\mathbf{Q} = -\lambda\operatorname{grad}\theta. \tag{3.4}$$

We do not describe in detail the chemical reactions that occur across the combustion front. In fact, a knowledge of the composition of the fluids in zones I and II suffices. Moreover, the coke phase no longer enters the problem because it is produced and destroyed in the reaction zone, whose width is neglected.

To describe the quantities that may exhibit discontinuities across the combustion front, we assume that the equations obtained in Section 1 are valid in the sense of distributions. We introduce $\psi^{\bullet}_\alpha$, the intensity of the reaction $\alpha$ per unit surface area of the reaction front $\psi_\alpha$, the unitary normal $\mathbf{n}$ on $\Sigma$ (the combustion front) in the direction from zone I to zone II, and the velocity $\mathbf{w}$ of the surface. in the previous equations we use instead of $\sum_\alpha \mathbf{v}_{i\alpha}\psi_\alpha$ a term of the form

$$\sum_\alpha \nu_{i\alpha}\,\psi^{\bullet}_\alpha\,[1 + (\mathbf{w.n})^2]^{-1/2}\,\delta_\Sigma\,,$$

where $\delta_\Sigma$ is the Dirac distribution of $\Sigma$ (regarded as a hypersurface in space and time).

In zones I and II the distribution $\delta_\Sigma$ vanishes, and thus Equations (3.1)–(3.4) are obtained.

From Equation (1.17) we deduce that $\theta$ is continuous across $\Sigma$, but its gradient can be discontinuous, as can the heat flux vector.

Similarly, from (1.15) the phase pressure $P_a$ is seen to be continuous, and clearly grad $P_a$ and $\mathbf{U}^a$ can be discontinous when across $\Sigma$.

Hence from (1.14) and (1.16) we get the equations

$$-\left[n\sum_a S_a R_a C_{ia}\right]\mathbf{w}\cdot\mathbf{n} + \left[\sum_a nS_a R_a C_{ia}\mathbf{U}^a\right]\cdot\mathbf{n} - \sum_a \nu_{ia}\psi^*_\alpha = 0\,, \tag{3.5}$$

$$-\left[n\sum_a S_a R_a E_a + (1-n)R_s E_s\right]\mathbf{w}\cdot\mathbf{n} + \left[\sum_a nS_a R_a h_a \mathbf{U}^a + \mathbf{Q}\right]\cdot\mathbf{n} = 0\,, \tag{3.6}$$

where $[\cdot]$ represents the jump on $\Sigma$, and hence the difference between the values on $\Sigma$ when passing from zone I to zone II.

Let us remark that in (3.5) and (3.6) the velocity $\mathbf{w}$ of the combustion front does not appear explicitly; only its normal component, which has a physical meaning, is present.

These conditions are not sufficient for solving the problem; other conditions, resulting from physical experiments, have to be added. A simple example will serve as an illustration. Let us consider a one-dimensional problem with five constitutents: oxygen, water, gas, and light and heavy hydrocarbons. A mixture of oxygen, water and gas is being injected. Hence there exist three fluid phases formed by gas (denoted by the subscript g), water (denoted by the subscript w) and oil (denoted by the subscript 'o', containing oxygen, water and the two kinds of hydrocarbon as constituents). There are only two chemical reactions as those described in the previous section. In Equations (3.5) and (3.6) appear the only component of the velocity of the combustion front and the only component of the seepage velocity.

We assume that zone I contains water and gas only, and that zone II contains all three fluid phases.

With these conditions the system (3.1)–(3.4) gives for zone I seven equations. To these we have to add the saturation and concentration equations $S_g + S_w = 1$, $\sum^3_{i=1} C_{ig} = 1$, the equation of local thermodynamic equilibrium (which yields the concentration of the water constituent from the gas phase in terms of the pressure in this phase, $P_g$, and $\theta$) as well as the equation for the capillary pressure $P_g - P_w = P_c(S_w)$. Hence we get 11 equations for 11 unknowns (three concentrations, two saturations, two pressures, two velocities, one temperature and the heat flux).

Similarly, the system (3.1)–(3.4) gives for zone II nine equations. The saturation and concentration equations are $S_g + S_w + S_o = 1$, $\sum\limits_{i=1}^{4} C_{ig} = 1$ and $\sum\limits_{i=4}^{5} C_{io} = 1$. The local thermodynamic equilibrium equations yield the concentration in the water constituent of the gas phase as a function of the pressure of this phase and of the temperature, and also the ratio between the concentrations of light hydrocarbons and temperature. By expressing the two capillary pressures as functions of saturation, we get 16 equations for the 16 unknowns (five concentrations, three saturations, three pressures, three velocities, one temperature and the heat flux).

For the combustion front the system given by (3.5) and (3.6) yields six equations. If we suppose that the gas in zone I does not contain light hydrocarbons, and that the oxygen is totally consumed, then on the surface we shall have as unknowns the values of the concentrations of oxygen, water and gas from the gas phase of zone I, the values of the concentrations in water, gas and light hydrocarbons from the gas phase of zone II, and the values of the cocentrations in light and heavy hydrocarbons from the oil phase of zone II. Hence there are eight unknowns, to which $\psi_1^*$, $\psi_2^*$, and $\mathbf{w}$ are added. It follows that to the six equations given by the system (3.5)−(3.6) we have to add three concentration equations and also two local thermodynamic equilibrium equations which express the concentrations in terms of the intensity of the chemical reactions. Thus the system of equations described above is completed.

A numerical method may be used, in which a solution is obtained at each time step of the systems of equations in zones I and II, and the velocity of the combustion front is computed in readiness for the next iteration.

## 4. Numerical Examples

Here we explain, using some simplifying assumptions, three phenomena that may be associated with flow through a porous medium in the presence of a temperature field. One is the importance of viscous dissipation in the energy balance. The second is the evolution of a temperature front with time, for which we take into account the loss of heat via conducting layers in the vicinity of the oil production layer. The third is the evolution of an intermediate zone, as in the case of the reaction zone, for which we must bear in mind that each zone can have its own pressure.

In each case we neglect any chemical reactions, but they are illustrative for a good understanding of physical situations that may arise in pr̆actical cases.

The general case of underground combustion may be treated using the general equations obtained in Section 1.

EXAMPLE 1. We consider the problem of heat propagation in one dimension. The velocity is known, and consequently we use only the equation (3.24) in Chapter 1 or (4.21) in Chapter 4:

$$\chi \Delta T - \rho_f C_f \mathbf{v} \operatorname{grad} T - [n \rho_f C_f + (1-n) \rho_s C_s] \frac{\partial T}{\partial t} = -\frac{\mu}{k} v^2,$$

where $T$ is the temperature, $\mathbf{v}$ the velocity, $k$ the permeability, $\mu$ the viscosity and

$\chi$ the thermal conductivity. This equation may be written for one dimension in the form

$$\chi \frac{\partial^2 T}{\partial x^2} - \rho_f C_f v \frac{\partial T}{\partial x} - [n\rho_f C_f + (1-n)\rho_s C_s] \frac{\partial T}{\partial t} = -\frac{\mu}{k} v^2. \tag{4.1}$$

The seepage velocity is given by Darcy's law as

$$v = \frac{k}{\mu} \frac{p_0 - p_L}{L} \tag{4.2}$$

where $L$ is the length of the medium and $p_0 - p_L$ is the pressure difference between $x = 0$ and $x = L$.

With the notation

$$\alpha = \frac{\rho_f C_f}{\chi} \frac{k}{\mu} \frac{p_0 - p_L}{L},$$

$$\beta = \frac{n\rho_f C_f + (1-n)\rho_s C_s}{\chi},$$

$$\gamma = \frac{1}{\chi} \frac{k}{\mu} \left( \frac{p_0 - p_L}{L} \right)^2.$$

Equation (4.1) takes the form:

$$\frac{\partial^2 T}{\partial x^2} - \alpha \frac{\partial T}{\partial x} - \beta \frac{\partial T}{\partial t} = -\gamma. \tag{4.3}$$

For this equation it is possible to formulate two initial and boundary-value problems.

The first is

$$T(x, 0) = 0, \qquad T(0, t) = 0, \qquad \frac{\partial T}{\partial x}(L, t) = 0, \tag{4.4}$$

and the second

$$T(x, 0) = 0, \qquad T(0, t) = 0, \qquad \frac{\partial T}{\partial x}(L, t) = -AT(L, t). \tag{4.5}$$

The problem (4.4) corresponds to the injection of a fluid at the same temperature as the initial temperature of the medium (cold fluid) and with zero normal heat flux at $x = L$. That means that at $x = L$ there is no loss of heat.

The problems (4.5) also corresponds to the injection of a cold fluid, but includes loss of heat at $x = L$. In this case at $x = L$ we may consider that there exists a production well. The proportionality factor $A$ in the third of the conditions (4.5) is of the form $A = \alpha n$.

For these problems it is possible to obtain analytical solutions, but the form of such solutions is very complicated and difficult for practical use. For this reason we shall give in detail a numerical solution.

We consider a discretization in space and time, with steps $h$ and $\tau$, respectively. The derivatives are defined by

$$\frac{\partial T}{\partial x}(x_m, t_p) = \frac{T(x_m, t_p + \tau) - T(x_m - h, t_p + \tau)}{h},$$

$$\frac{\partial^2 T}{\partial x^2}(x_m, t_p) = \frac{T(x_m - h, t_p) - 2T(x_m, t_p) + T(x_m + h, t_p)}{h^2},$$

$$\frac{\partial T}{\partial t}(x_m, t_p) = \frac{T(x_m, t_p + \tau) - T(x_m, t_p)}{\tau}.$$

If we denote by $x_m = (m-1)\,h$, $t_p = (p-1)\,\tau$ and $T_{m,p} = T(x_m, t_p)$, Equation (4.3) takes the form

$$\frac{T_{m-1,p} - 2\,T_{m,p} + T_{m+1,p}}{h^2} - \alpha\,\frac{T_{m,p+1} - T_{m-1,p+1}}{h} - \beta\,\frac{T_{m,p+1} - T_{m,p}}{\tau} = -\gamma$$

or, equivalently,

$$T_{m,p+1} = \frac{\gamma\tau + \alpha h R T_{m-1,p+1} + (\beta - 2R)\,T_{m,p} + R\,(T_{m-1,p} + T_{m+1,p})}{\beta + \alpha h R}, \qquad (4.6)$$

where $R = \dfrac{\tau}{h^2}$, $m = 2, 3, \ldots, M-1$, $p = 1, 2, \ldots$, where $M = \left[\dfrac{L}{h}\right] + 1$. The form of discretization used above is shown in Fig. 27.

The initial and boundary conditions (4.4) and (4.5) are now

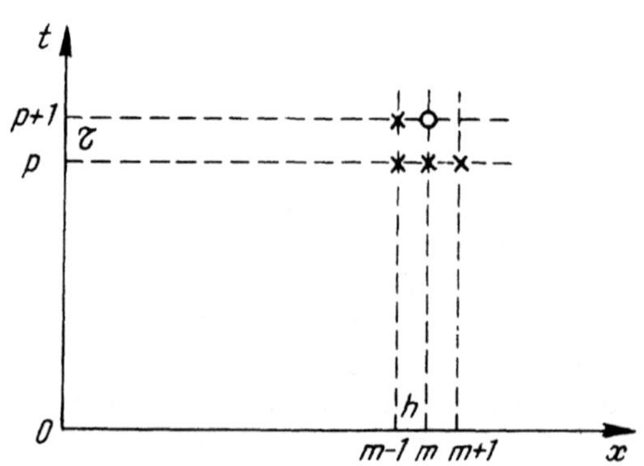


Fig. 27. Discretization scheme.

$$\begin{aligned}
&T_{m,1} = 0, && m = 1, 2, \ldots, \mathrm{M},\\
&T_{1,p} = 0, && p = 1, 2, \ldots,\\
&T_{M,p} = T_{M-1,p}, && p = 1, 2, \ldots,
\end{aligned} \tag{4.7}$$

$$\begin{aligned}
&T_{m,1} = 0, && m = 1, 2, \ldots, M,\\
&T_{1,p} = 0, && p = 1, 2, \ldots,\\
&(1+Ah)T_{M,p} = T_{M-1,p}, && p = 1, 2, \ldots .
\end{aligned} \tag{4.8}$$

With this method we can determine the values of temperature $T_{m,p}$, step by step, at any point $x_m$ and any time $t_p$. The first condition in (4.7) or (4.8) gives the temperature for $p = 1$ and $m = 1, 2, \ldots, M$, and the second condition of (4.7) or (4.8) gives the temperature for $p = 2$, $T_{1,2}$, and then using (4.6) for $p = 1$ and $m = 2, 3, \ldots, M-1$, we obtain the temperature at the time $\tau$ and at the points $x_2, x_3, \ldots, x_{M-1}$. Then, for $p = 2$ we have $T_{M,2}$ from the third condition in (4.7) or (4.8). Similarly we obtain the temperature at the time steps $2\tau, 3\tau, \ldots$.

The problem of the stability of this discretization scheme may be solved by using spectral analysis. If we search for a solution of the form $T_{m,p} = \lambda^p \exp(im\varphi)$, it is known that the stability condition is expressed by $|\lambda(\varphi)| \leqslant 1$. For Equation (4.6) we obtain

$$(\alpha^2 h^2 - 4\sin^2 \frac{1}{2}\varphi)\, R + \beta\,(\tau h + 2) \geqslant 0\,.$$

Then there are two possibilities: if $\alpha^2 h^2 - 4\sin^2 \frac{1}{2}\varphi \geqslant 0$, we obtain $h < 2/\alpha$, and this condition is satisfied for every $\tau$; if $\alpha^2 h^2 - 4\sin^2 \frac{1}{2}\varphi < 0$, the stability condition is $R < \beta/2$.

Let us choose the values $h = 0.1$ m, $\tau = 0.5$ h and $R = 50 < \beta/2$. If we consider injection with water, we have for the physical constants the values: $\rho_f = 10^3$ kg m$^{-3}$, $C_f = 1$ kcal kg$^{-1}$ K$^{-1}$ and $\mu = 0.5055 \times 10^{-7}$ kg h$^{-1}$ m$^{-1}$. If we take the solid skeleton to be sandstone, as is usual in oil fields, then $\rho_s C_s = 480.519$ kcal m$^{-3}$ K$^{-1}$ and $\chi = 1.3394$, kcal m$^{-1}$ h$^{-1}$ K$^{-1}$.

The first computation is made for a permeability $k = 1$ D, a porosity $n = 0.4$ and a porous medium 100 m in length, in order to observe the difference between the solutions of the problems (4.7) and (4.8). This difference is represented in Fig. 28. It is easy to see that for Problem (4.7) the value of the temperature is constant at 5.0788°C for $t = 126$ h, while for Problem (4.8) at $L = 99.8$ m the temperature starts to fall, and at $L = 100$ m the value is 0.2029°C. This computation shows us that the loss of heat is important only in the neighbourhood of the production well, and consequently the results will be practically the same for both boundary-value problems. For this reason, we restrict ourselves to Problem (4.7).

We now consider the heat flow in the porous medium. If we take a rate of fluid flow $Q = 0.1928$ m$^3$h$^{-1}$ at $p = 1$ atm and for $L = 50$ m, we obtain for different values of time the curves shown in Fig. 29. Generally the increase in temperature is linear up to a certain value (depending on time), and a distance x,

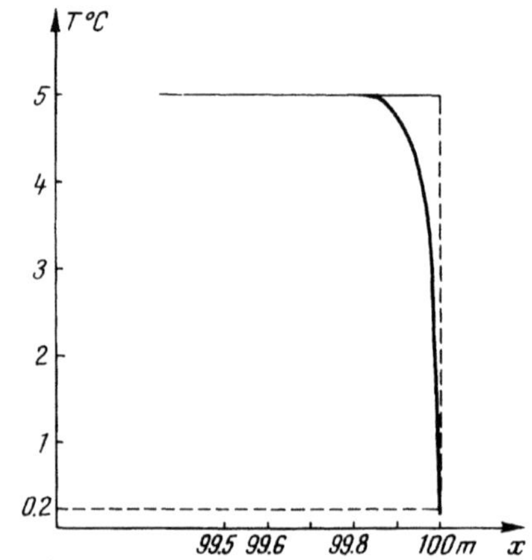


Fig. 28. The loss of heat at $x = L$.

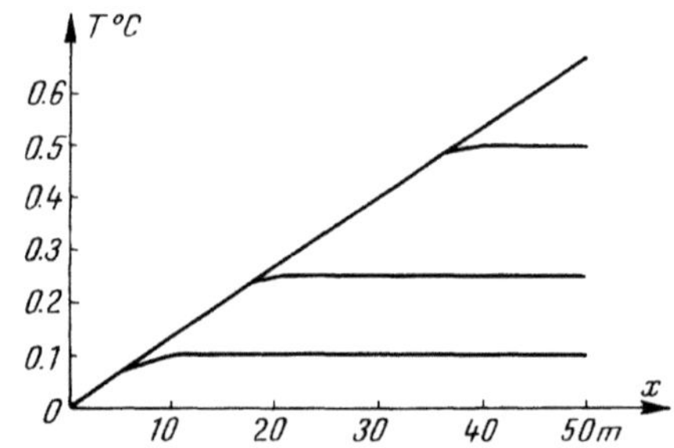


Fig. 29. Heat propagation for different times.

after which it remains constant. This phenomenon shows that at a distance $x_0$ it is necessary to impose the last condition in (4.7). This condition give us a stabilization of the heat at a value depending on $x_0$ and time. The extremity $x = L$ imposes a maximum value for the temperature.

The influence of the length of the porous medium is shown in Fig. 30. The values obtained for $L = 10, 50, 100$, and 150m are $T_{10} = 0.1257$°C (at $t_{10} = 39$ h), $T_{50} = 0.6341$°C (at $t_{50} = 230$h), $T_{100} = 1.2695$°C (at $t_{100} = 418$ h) and $T_{150} = 1.9048$°C (at $t_{150} = 603$h).

The pressure gradient has a great influence on the production of heat by viscous dissipation. The viscous dissipation $\gamma$ is proportional to the square of the pressure gradient, and consequently at high values of this gradient it is possible to obtain high values for the temperature. Figures 31 – 34 show the variation of temperature with $x$ plotted at different times, for $p = 1, 2, 3$, and 4 atm. Note that after approximately 25 days the temperature remains constant for all values of the pressure gradient.

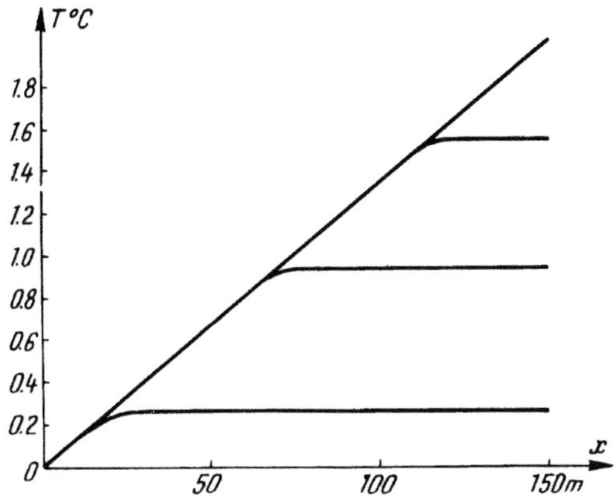


Fig. 30. Temperature as function of time and length of medium.

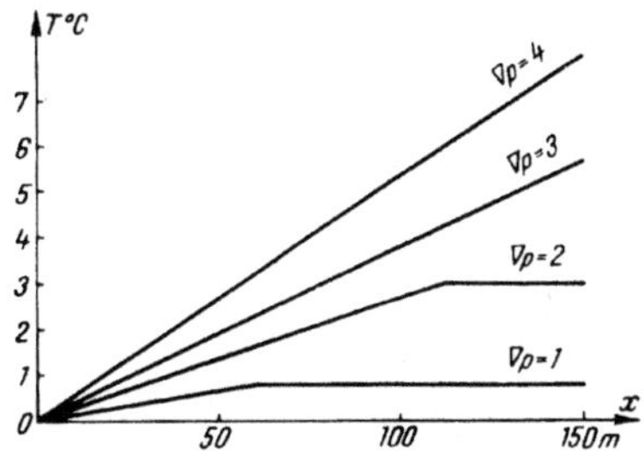


Fig. 31. Temperature at $t = 8$ days.

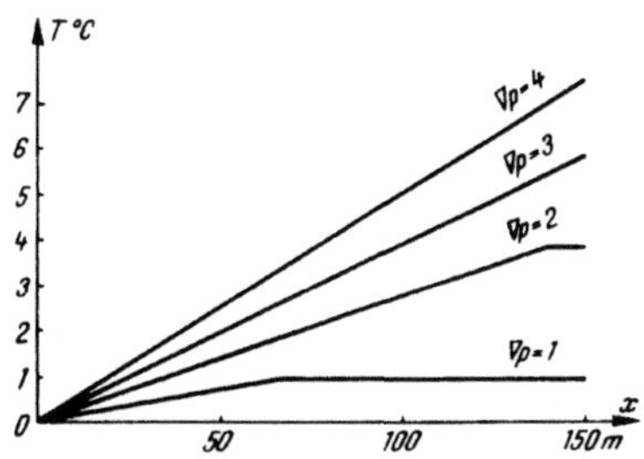


Fig. 32. Temperature at $t = 10$ days.

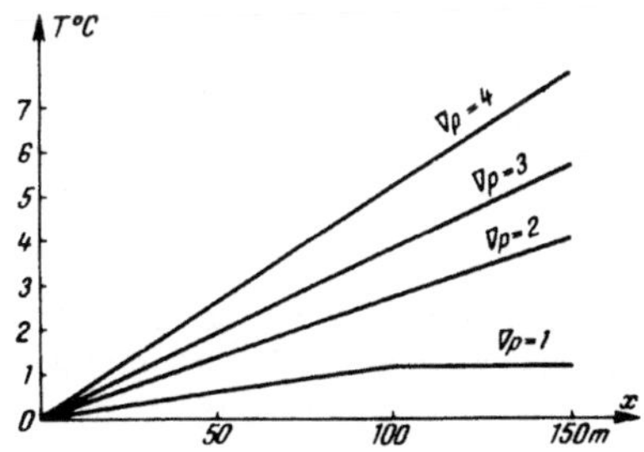


Fig. 33. Temperature at $t = 14$ days.

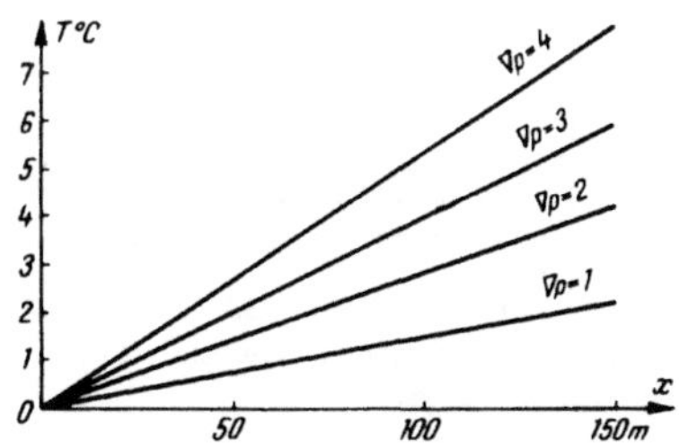


Fig. 34. Temperature at $t = 25$ days.

The permeability of the porous medium also has an important effect on the increase of temperature since the viscous dissipation and convection terms are proportional to $k$. Figures 35 – 38 show the increase of temperature plotted for differnet values of $k$. These plots are much the same as those of the pressure gradient.

We now consider the effects of viscosity, and use a value $C_f = 0.5$ kcal kg$^{-1}$ K$^{-1}$

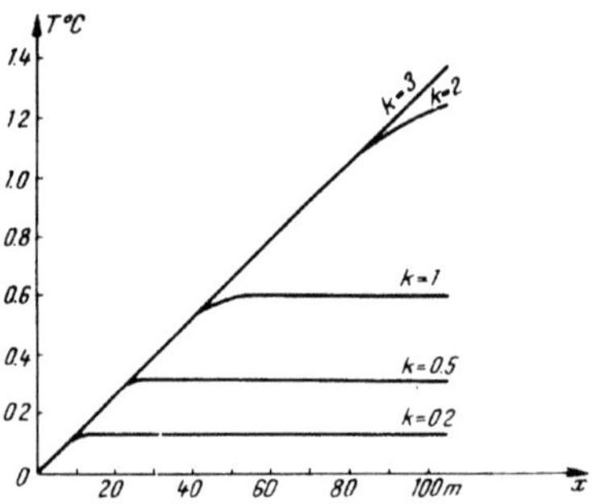


Fig. 35. Increase of temperature for different $k$, at $t = 7$ days.

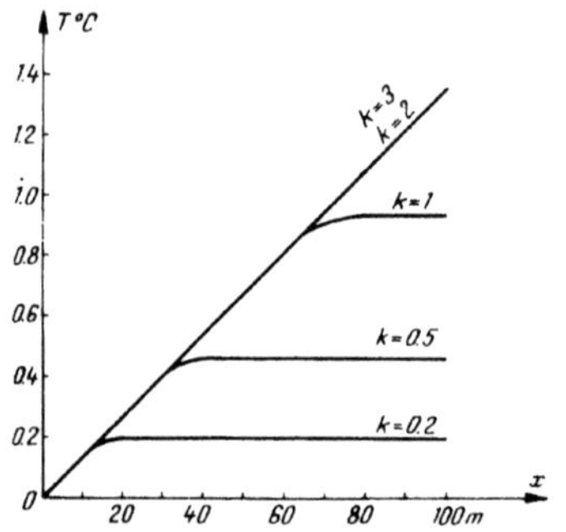


Fig. 36. Increase of temperature for different $k$, at $t = 11$ days.

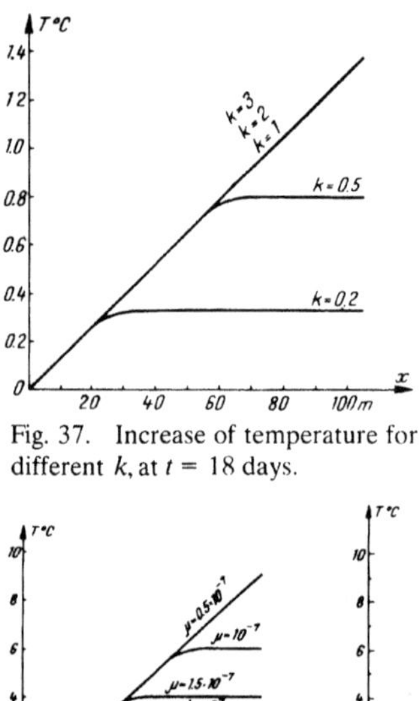


Fig. 37. Increase of temperature for different $k$, at $t = 18$ days.

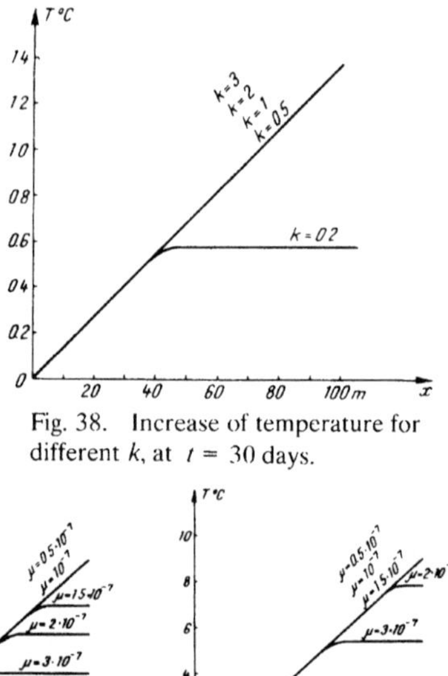


Fig. 38. Increase of temperature for different $k$, at $t = 30$ days.

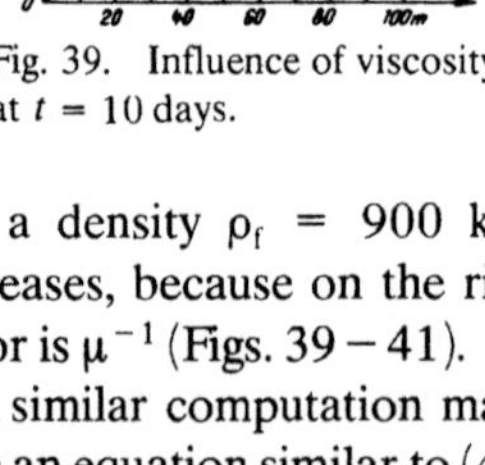


Fig. 39. Influence of viscosity at $t = 10$ days.

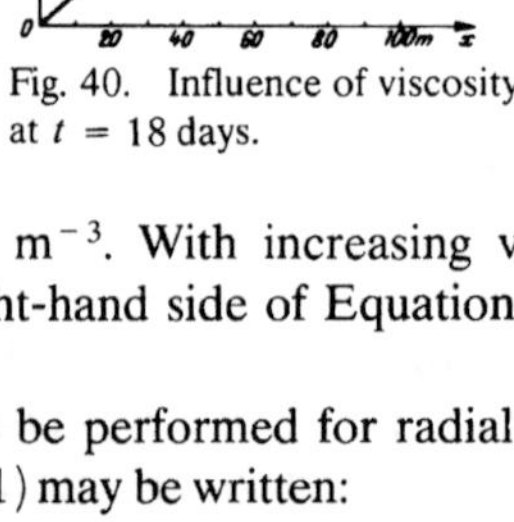


Fig. 40. Influence of viscosity at $t = 18$ days.

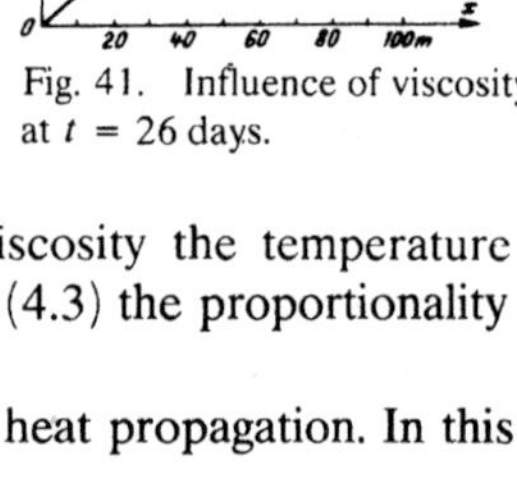


Fig. 41. Influence of viscosity at $t = 26$ days.

and a density $\rho_f = 900 \text{ kg m}^{-3}$. With increasing viscosity the temperature decreases, because on the right-hand side of Equation (4.3) the proportionality factor is $\mu^{-1}$ (Figs. 39–41).

A similar computation may be performed for radial heat propagation. In this case an equation similar to (4.1) may be written:

$$\chi\left(\frac{\partial^2 T}{\partial r^2} + \frac{1}{r}\frac{\partial T}{\partial r}\right) - \rho_f C_f v \frac{\partial T}{\partial r} + \frac{\mu}{k} v^2 = [n\rho_f C_f + (1-n)\rho_s C_s]\frac{\partial T}{\partial t}, \qquad (4.9)$$

where $v = Q/2\pi r$ is the radial velocity. If we introduce the notation

$$\alpha_1 = \frac{\rho_f C_f Q}{4\pi\chi},$$

$$\beta = \frac{n\rho_f C_f + (1-n)\rho_s C_s}{\chi},$$

$$f(r) = \frac{\mu Q^2}{4\pi^2 k\chi}\,\frac{1}{r^2}.$$

Equation (4.9) takes the form

$$\frac{\partial^2 T}{\partial r^2} + (1-2\alpha_1)\frac{1}{r}\frac{\partial T}{\partial r} - \beta\frac{\partial T}{\partial t} = -f(r). \tag{4.10}$$

If $r = r_w$ is the equation for the injection well, the initial and boundary conditions are

$$T(r, 0) = 0, \qquad T(r_w, t) = 0. \tag{4.11}$$

We consider this problem for $r \in (r_w, +\infty)$, so we add the condition that the solution will be bounded in the whole space.

For the numerical computation we use a discretization (Fig. 42) of the form $r_m = r_w + (m-1)h$ $(m = 1, 2, \ldots)$, $t_p = (p-1)\tau$ $(p = 1, 2, \ldots)$, $T_{m,p} = T(r_m, t_p)$, and the derivatives

$$\frac{\partial T}{\partial r} = \frac{T_{m,p+1} - T_{m-1,p+1}}{h},$$

$$\frac{\partial^2 T}{\partial r^2} = \frac{T_{m-1,p} - 2T_{m,p} + T_{m+1,p}}{h^2},$$

$$\frac{\partial T}{\partial t} = \frac{T_{m,p+1} - T_{m,p}}{\tau}.$$

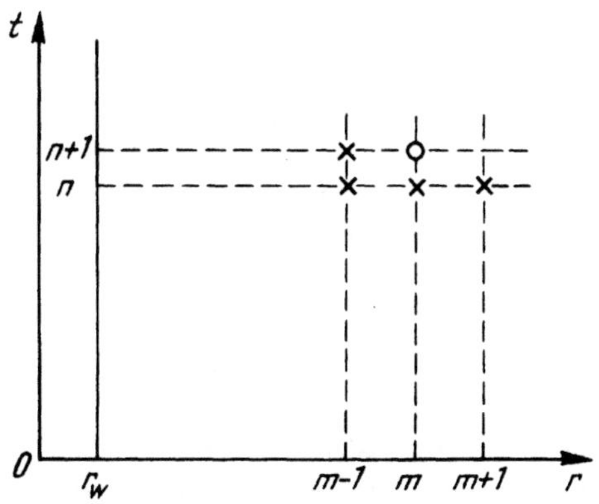


Fig. 42. The finite-difference scheme.

Then Equation (4.10) takes the form

$$\frac{T_{m-1,p} - 2T_{m,p} + T_{m+1,p}}{h^2} + \frac{1-2\alpha_1}{(r_w/h) + m - 1} \frac{T_{m,p+1} - T_{m-1,p+1}}{h} -$$

$$- \beta \frac{T_{m,p+1} - T_{m,p}}{\tau} = - \frac{\gamma_1}{h^2((r_w/h) + m - 1)^2},$$

$$\gamma_1 = \frac{\mu Q^2}{4\pi^2 k \chi}, \tag{4.12}$$

and the conditions (4.11) become

$$T_{m,1} = 0, \quad T_{1,p} = 0, \quad m, p = 1, 2, \ldots . \tag{4.13}$$

The stability condition for this problem also leads to two possibilities: if $(2\alpha_1 - 1)/((r_w/h) + m - 1) > 2$ the finite-difference scheme is stable for any $h$ and $\tau$ and if $R < \frac{\beta}{2}$ the scheme is also stable. This last condition is our stability condition.

The physical constants used are the same as in the linear case. From Figs. 43–45 we can show that the injection time is important only at the beginning. For example, from Fig. 43 the temperature is seen to increase to a maximum of 8.171°C after 10 days and a maximum of 9.246°C after 30 days. Similarly, Fig. 44 shows that after 10 days the maximum temperature is 10.341°C, and after 30 days 18.488°C. These results are very important from the practical point of view: they show that the injection time necessary to achieve a certain level of temperature can be reduced, due to the viscous-dissipation.

As in the linear problem, the permeability of the medium is a very important parameter. For a high permeability, the increase in temperature is negligibly small, but for a small permeability it is possible to obtain high temperatures. For example for $k = 0.003$ D and after 15 days of injection, we obtain a maximum temperature of 28.5406°C (Figs. 46 and 47).

The importance of the other parameters is much the same as in the linear problem.

It is important to note that the effect of the heat produced by viscous dissipation is important only for spontaneous ignition of the oil field. In the general case of underground combustion when the temperature is of the order of hundreds of degrees Celsius, it is clear that an increase with of a few degrees is negligible.

EXAMPLE 2. We consider a temperature front propagating radially. In order to write temperature equations for the upstream and downstream zones, we look for the balance of energy in a domain $D = \{a \leqslant r \leqslant b, \quad h_1 \leqslant z \leqslant h_2\}$ in the variables $r$ and $z$. It is possible to prove that this balance equation is independent of $a$ and $b$, but that it depends on the normal heat flux through the boundaries

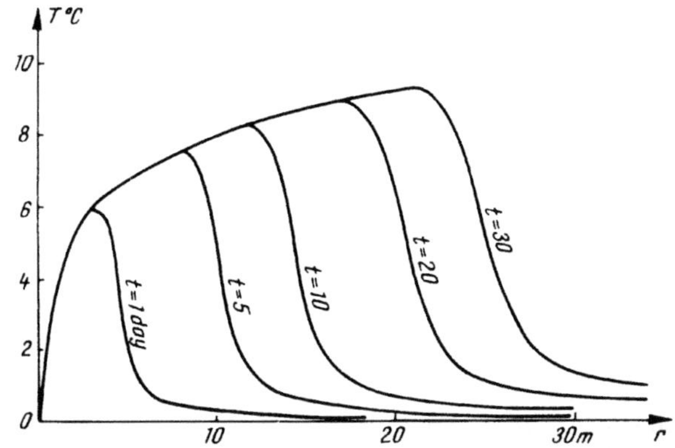

Fig. 43. Temperature after different times, for $k = 0.01$ D.

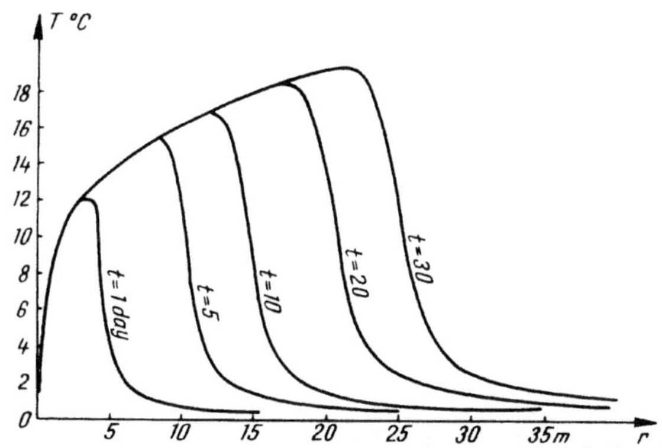

Fig. 44. Temperature after different times, for $k = 0.005$ D.

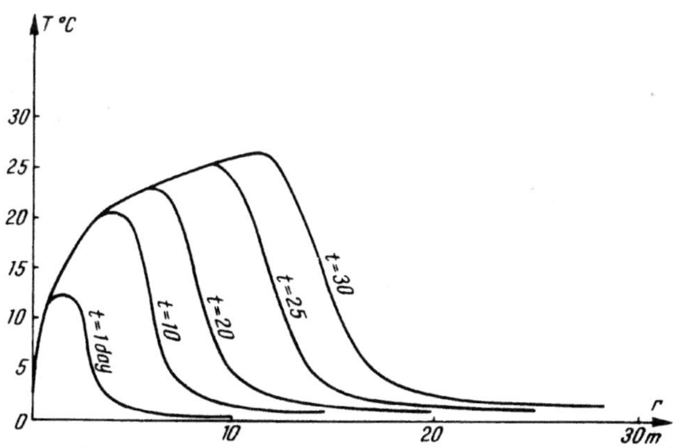

Fig. 45. Temperature after different times, for $\mu = 1.5 \times 10^{-7}$ kg h$^{-1}$m$^{-1}$ and $k = 0.01$ D.

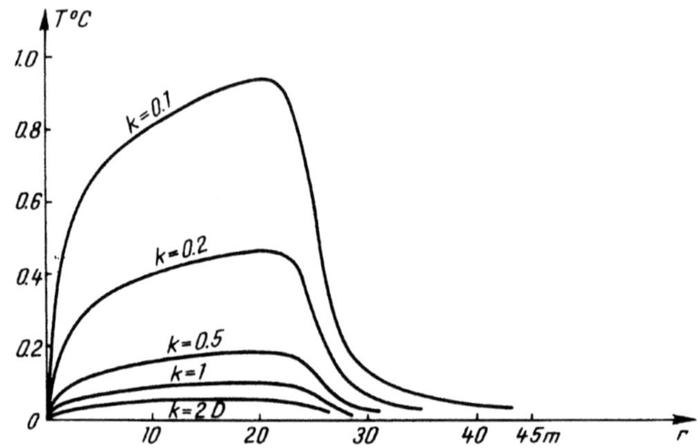


Fig. 46. Temperature for different permeabilities, at $t$= 30 days.

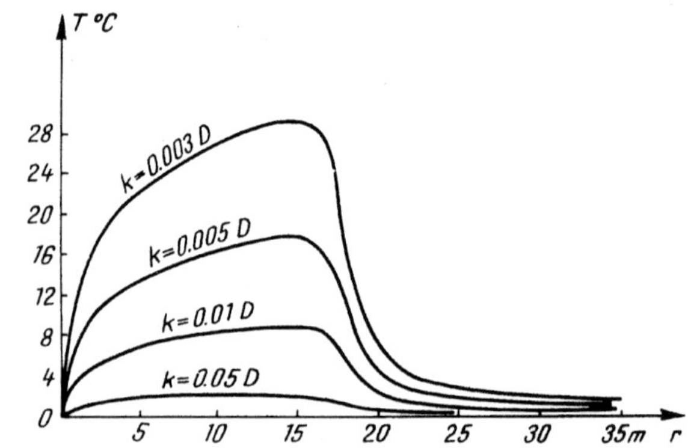


Fig. 47. Temperature for different permeabilities, at $t$= 15 days.

$z = h_1$ and $z = h_2$. If we neglect the lose of heat by conduction at $z = = h_1$ and $z = = h_2$, the equation is similar to (4.21) in Chapter 4, but if we take this loss of heat into consideration, terms containing the normal heat flux at $z = h_1$ and $z = h_2$ will appear in the equation.

We therefore consider the following problem:

$$\chi_1 \frac{\partial^2 \theta^1}{\partial z^2} = C_1 \frac{\partial \theta^1}{\partial t} \quad \text{in } D_1 = \{h_2 < z < H, 0 < t < T_0\}, \tag{4.14}$$

$$\chi_1^* \left( \frac{\partial^2 T^1}{\partial r^2} + \frac{1}{r} \frac{\partial T^1}{\partial r} \right) + \frac{\chi_1}{H} \frac{\partial \theta^1}{\partial z}\bigg|_{z=h_2} - \frac{\chi_2}{H} \frac{\partial \theta^2}{\partial z}\bigg|_{z=h_1} =$$

$$= C_1^* \left( \frac{\partial T^1}{\partial t} + v_1 \frac{\partial T^1}{\partial r} \right) \quad \text{in } D^1 = \{r_w < r < R(t),\ 0 < t < T_0\}, \tag{4.15}$$

$$\chi_2^* \left( \left( \frac{\partial^2 T^2}{\partial r^2} + \frac{1}{r}\frac{\partial T^2}{\partial r} \right) + \frac{\chi_1}{H}\frac{\partial \theta^1}{\partial z}\bigg|_{z=h_2} - \frac{\chi_2}{H}\frac{\partial \theta^2}{\partial z}\bigg|_{z=h_1} = \right. \tag{4.16}$$

$$= \dot{C}_2^* \left( \frac{\partial T^2}{\partial t} + v_2 \frac{\partial T^2}{\partial r} \right) \quad \text{in } D^2 = \{R(t) < r < L, 0 < t < T_0\},$$

$$\chi_2 \frac{\partial^2 \theta^2}{\partial z^2} = C_2 \frac{\partial \theta^2}{\partial t} \quad \text{in } D_2 = \{0 < z < h_1, 0 < t < T_0\}, \tag{4.17}$$

with the initial and boundary conditions

$$\theta^1(r, z, 0) = f_1, \tag{4.18}$$

$$\theta^2(r, z, 0) = f_2, \tag{4.19}$$

$$\theta^1(r, H, t) = f_1, \quad r_w < r < L, \tag{4.20}$$

$$\theta^2(r, 0, t) = f_2, \quad r_w < r < L, \tag{4.21}$$

$$T^1(r, t) = \theta^1(r, h_2, t), \quad r_w < r \leqslant R(t), \tag{4.22}$$

$$T^2(r, t) = \theta^1(r, h_2, t), \quad R(t) \leqslant r < L, \tag{4.23}$$

$$T^1(r, t) = \theta^2(r, h_1, t), \quad r_w < r \leqslant R(t), \tag{4.24}$$

$$T^2(r, t) = \theta^2(r, h_1, t), \quad R(t) \leqslant r < L, \tag{4.25}$$

$$T^i(r, 0) = g_i, \quad i = 1, 2, \tag{4.26}$$

$$T^1(R(t), t) = T^2(R(t), t) = T(R(t), t), \tag{4.27}$$

$$T(R(t), t) = f(\ln P), \tag{4.28}$$

$$T^1(r_w, t) = \psi_1, \tag{4.29}$$

$$T^2(L, t) = \psi_2, \tag{4.30}$$

where $T^1$ is the temperature in the upstream zone, $T^2$ is the temperature in the downstream zone, $\theta^1$ and $\theta^2$ are the temperatures in the two layers in which there is only conduction (Fig. 48), $T$ is the temperature at the front $r = R(t)$, $P$ is the pressure on the front, and $v_1$ and $v_2$ are the velocities in the two domains. In this problem $v_1$ and $v_2$ are functions of the rate of flow and of $r$, and are of the same form as in the radial problem treated in Example 1:

$$v_1 = \frac{Qc_1}{2\pi H\gamma_1 C_1^* r}, \qquad v_2 = \frac{Qc_2}{2\pi H\gamma_2 C_2^* r},$$

where $c_1$ and $c_2$ are the heat capacities of the hot injection fluid and of the extracted fluid.

In this problem, which is in fact a Stephan problem, the equation for the combustion front has the form

$$\frac{\mathrm{d}R}{\mathrm{d}t} = A\left[\chi_2^* \frac{\partial T^2}{\partial r}\bigg|_{r=R(t)} - \chi_1^* \frac{\partial T^1}{\partial r}\bigg|_{r=R(t)} + B\frac{T(R(t), t)}{R(t)}\right], \tag{4.31}$$

Fig. 48.

where $A$ and $B$ depend on the physical properties of each fluid phase, and on $P(R(t), t)$. The pressure $P$ on the front is calculated from the formula

$$P = P_* + \frac{Q\mu}{2\pi k H \gamma_2} \ln \frac{R_*}{R(t)}, \tag{4.32}$$

where $k$ is the permeability of the medium, $\mu$ is a mean value of the viscosity of the fluid phase, $R_*$ is the distance between the injection and extraction wells and $P_*$ is the pressure at $x = L$.

The temperature at the combustion front is taken as function of $\ln P$ in the form

$$T = \frac{a}{(\ln P) - b} - 273\,.$$

The numerical computation of this problem is performed by a finite-difference scheme in dimensionless form (for details of which see Abatuliev et al. (1978)). We simply indicate here the values of the parameters used in the computation: $\chi_1^* = 0.1$ kcal m$^{-1}$ h$^{-1}$ K$^{-1}$, $\chi_2^* = 1.5$ kcal m$^{-1}$ h$^{-1}$ K$^{-1}$, $C_1^* = 1.5$ kcal m$^{-3}$ K$^{-1}$, $C_2^* = 500$ kcal m$^{-3}$ K$^{-1}$, $\gamma_1 = 5$ kg m$^{-3}$, $\gamma_2 = 1000$ kg m$^{-3}$, $c_1 = 4$ kcal m$^{-3}$ K$^{-1}$, $c_2 = 1400$ kcal m$^{-3}$ K$^{-1}$, $k = 0.1$ D, $\chi_1 = \chi_2 = 0{,}86$ kcal m$^{-1}$ h$^{-1}$ K$^{-1}$, $C_1 = 500$ kcal m$^{-3}$ K$^{-1}$, $C_2 = 150$ kcal m$^{-3}$ K$^{-1}$, $P_* = 50$ atm, $R_* = 100$ m, Q = 2000 or 3000 kg h$^{-1}$.

Fig. 49 shows the difference between the solution of this problem, and the one in which the loss of heat is neglected, treated by Abasov (1972). The differences are important, and they increase with time.

EXAMPLE 3. It is well known that a porous medium may be deformed under the pressure gradient applied in oil fields. Here we shall try to explain the so-called Tchearni equation (Tchearni 1948), used for deformable porous media. For this, we take Equations (4.16) and (4.19) from Chapter 4 and introduce the seepage velocity $v_k = nu_k$:

$$\frac{\partial}{\partial t}(n\rho) + \frac{\partial}{\partial x_k}(\rho v_k) = 0\,, \tag{4.33}$$

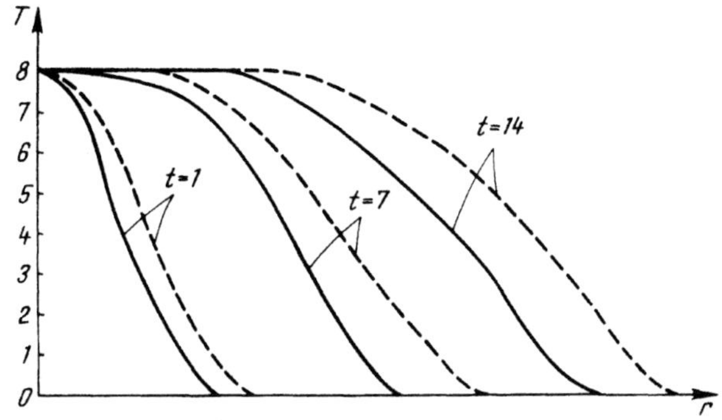


Fig. 49.

$$v_k = -\frac{k}{\mu}\frac{\partial p}{\partial x_k} - \frac{k\rho}{n\mu}\frac{\partial v_k}{\partial t}, \tag{4.34}$$

where the exterior body forces were omitted. We take the compressibility equations for the fluid and solid phases,

$$\rho = \rho_0\left[1 + b_f(p - p_0)\right],$$

$$n = n_0 + b_s(p - p_0),$$

and, neglecting the terms containing $(p-p_0)^2$ of the product $\rho n$, we have

$$\rho n = \rho_0 n_0 + \rho_0 (p - p_0)(b_s + n_0 b_f).$$

If in (4.33) the product $v_k \dfrac{\partial \rho}{\partial x_k}$ is also neglected, then we have

$$\rho_0(b_s + n_0 b_f)\frac{\partial p}{\partial t} = -\rho_0 \frac{\partial v_k}{\partial x_k}. \tag{4.35}$$

Applying the divergence operator to Equation (4.34), we obtain

$$\frac{\partial v_k}{\partial x_k} = -\frac{k}{\mu}\frac{\partial^2 p}{\partial x_k \partial x_k} - \frac{k\rho_0}{\mu n_0}\frac{\partial^2 v_k}{\partial t\, \partial x_k}. \tag{4.36}$$

From (4.35) and (4.36) we have, successively:

$$(b_s + n_0 b_f)\frac{\partial p}{\partial t} = \frac{k}{\mu}\frac{\partial^2 p}{\partial x_k \partial x_k} - \frac{k\rho_0}{\mu n_0}(b_s + n_0 b_f)\frac{\partial^2 p}{\partial t^2},$$

$$\Delta p = (b_s + n_0 b_f)\left(\frac{\mu}{k}\frac{\partial p}{\partial t} + \frac{\rho_0}{n_0}\frac{\partial^2 p}{\partial t^2}\right),$$

$$\Delta = \frac{\partial^2}{\partial x^2} + \frac{\partial^2}{\partial y^2} + \frac{\partial^2}{\partial z^2}.$$

Introducing $n_0 K^{-1} = b_s + n_0 b_f$ and $c^{-2} = \rho_0 K^{-1}$, we have

$$\Delta p = \frac{1}{c^2}\frac{\partial^2 p}{\partial t^2} + \frac{\mu n_0}{kK}\frac{\partial p}{\partial t} \tag{4.37}$$

Using a characteristic length $L$ of the medium and a characteristic pressure $p_0$, the dimensionless form of (4.37) is

$$\Delta p^* = \frac{\partial^2 p^*}{\partial \tau^2} + A\frac{\partial p^*}{\partial \tau}, \tag{4.38}$$

where $p^* = pp_0$, $\tau = ct/L$ and $A = \mu n_0 CL/kK$. But Tchearni (1948) observes that $A \gg 1$ in practice, and consequently (4.38) is of the form

$$\Delta p^* = A\frac{\partial p^*}{\partial \tau}$$

or, equivalently

$$\Delta p = \frac{1}{\varkappa}\frac{\partial p}{\partial t}, \quad \varkappa = \frac{kK}{\mu n_0}, \tag{4.39}$$

$\varkappa$ is a conductibility coefficient.

We can now formulate the problem of the evolution of an intermediate zone, in a deformed porous medium, of the form (4.39). For that we consider radial heat propagation, with the domain formed by $D_1$: $0 \leqslant r \leqslant r_1(t)$, $D_2$: $r_1(t) \leqslant r \leqslant r_2(t)$ and $D_3$: $r_2(t) \leqslant r < \infty$, and the systems of equations

$$\frac{\partial^2 p_i}{\partial r^2} + \frac{1}{r}\frac{\partial p_i}{\partial r} = \frac{1}{\varkappa_i}\frac{\partial p_i}{\partial t}, \qquad i = 1, 2, 3, \tag{4.40}$$

$$a_i\left(\frac{\partial^2 T_i}{\partial r^2} + \frac{1}{r}\frac{\partial T_i}{\partial r}\right) - v_i\frac{\partial T_i}{\partial r} = \frac{\partial T_i}{\partial t}, \qquad i = 1, 2, 3, \tag{4.41}$$

$$v_i = -\frac{kk_i}{\mu_i}\frac{\partial p_i}{\partial r}, \qquad i = 1, 2, 3, \tag{4.42}$$

with the initial and boundary conditions

$$p_3(r, 0) = p_0, \quad T_3(r, 0) = T_0, \tag{4.43}$$

$$\lim_{r \to 0}\left(r\frac{\partial p_1}{\partial r}\right) = -\frac{Q\mu_1}{2\pi kh}, \quad T_1(0, t) = T_w, \tag{4.44}$$

$$p_1(r_1, t) = p_2(r_1, t), \quad T_1(r_1, t) = T_2(r_1, t) = T^*, \tag{4.45}$$

$$\frac{k_1}{\mu_1}\frac{\partial p_1(r, t)}{\partial r} = \frac{k_2}{\mu_2}\frac{\partial p_2(r_1, t)}{\partial r} \tag{4.46}$$

$$q_1 \frac{dr_1}{dt} = \chi_1 \frac{\partial T_1(r_1, t)}{\partial r} - \chi_2 \frac{\partial T_2(r_1, t)}{\partial r}, \tag{4.47}$$

$$p_2(r_2, t) = p_3(r_2, t), \qquad T_2(r_2, t) = T_3(r_2, t) = T^{**}, \tag{4.48}$$

$$\frac{k_2}{\mu_2} \frac{\partial p_2(r, t)}{\partial r} = \frac{k_3}{\mu_3} \frac{\partial p_3(r_3, t)}{\partial r}, \tag{4.49}$$

$$q_2 \frac{dr_2}{dt} = \chi_2 \frac{\partial T_2(r_2, t)}{\partial r} - \chi_3 \frac{\partial T_3(r_3, t)}{\partial r}, \tag{4.50}$$

$$p_3(\infty, t) = p_0, \qquad T_3(\infty, t) = T_0 . \tag{4.51}$$

Note that the evolution of contact lines is described by Equations (4.47) and (4.50), which are of the form (4.31). That means that the velocity of such a line is proportional to the difference between the normal fluxes in the two regions.

The solutions of the system (4.40) may be obtained in the explicit form

$$p_i = A_i \operatorname{Ei}(-\xi_i) + B_i, \quad i = 1, 2, 3, \tag{4.52}$$

where

$$-\operatorname{Ei}(-x) = \int_x^\infty \frac{e^{-\xi}}{\xi} d\xi \cdot$$

Using (4.43)–(4.51), we obtain the coefficients

$$A_1 = -\frac{Q\mu_1}{4\pi kh}, \tag{4.53}$$

$$A_2 = -\frac{Q\mu_2}{4\pi kh} \exp\left[\nu_1^2 \left(\frac{1}{4\varkappa_2} - \frac{1}{4\varkappa_1}\right)\right], \tag{4.54}$$

$$A_3 = -\frac{Q\mu_3}{4\pi kh} \exp\left[\nu_1^2 \left(\frac{1}{4\varkappa_2} - \frac{1}{4\varkappa_1}\right) + \nu_2^2 \left(\frac{1}{4\varkappa_3} - \frac{1}{4\varkappa_2}\right)\right], \tag{4.55}$$

$$B_3 = p_0, \tag{4.56}$$

$$B_2 = A_3 \operatorname{Ei}\left(-\frac{\nu_2^2}{4\varkappa_3}\right) - A_2 \operatorname{Ei}\left(-\frac{\nu_2^2}{4\varkappa_2}\right) + P_0, \tag{4.57}$$

$$B_1 = A_2 \operatorname{Ei}\left(-\frac{\nu_1^2}{4\varkappa_2}\right) - A_1 \operatorname{Ei}\left(-\frac{\nu_1^2}{4\varkappa_1}\right) + B_2, \tag{4.58}$$

$\nu_1$ and $\nu_2$ being the velocities of the separation lines $r_1 = \nu_1 t$ and $r_2 = \nu_2 t$.

Then, using (4.52) and (4.42), the system (4.41) becomes

$$\frac{d^2 T_i}{d\xi_i^2} = -\left(\frac{1}{\xi_i} + \alpha_i \frac{e^{-\xi_i}}{\xi_i} + \beta_i\right) \frac{dT_i}{d\xi_i}, \quad i = 1, 2, 3, \tag{4.59}$$

with

$$\xi_i = \frac{r^2}{4\varkappa_i t}, \qquad \alpha_i = \frac{A_i k k_i}{\mu_i a_i}, \qquad \beta_i = \frac{\varkappa_i}{a_i}, \qquad i = 1, 2, 3.$$

The solution of (4.59) is

$$T_i(\xi_i) = M_i \varphi_i(\alpha_i, \beta_i, \xi_i) + N_i, \quad i = 1, 2, 3, \tag{4.60}$$

$$\varphi(\alpha_i, \beta_i, \xi_i) = \int_{\xi_i}^{\infty} \exp[-\alpha_i \mathrm{Ei}(-x) - \beta_i x] \frac{dx}{x} . \tag{4.61}$$

From (4.43)–(4.45), (4.48) and (4.51), we obtain

$$M_1 = \frac{T_w - T^*}{\varphi(\alpha_1, \beta_1, 0) - \varphi\left(\alpha_1, \beta_1, \frac{v_1^2}{4\varkappa_1}\right)}, \tag{4.62}$$

$$M_2 = \frac{T^* - T^{**}}{\varphi\left(\alpha_2, \beta_2, \frac{v_1^2}{4\varkappa_2}\right) - \varphi\left(\alpha_2, \beta_2, \frac{v_2^2}{4\varkappa_2}\right)}, \tag{4.63}$$

$$M_3 = \frac{T^{**} - T_0}{\varphi\left(\alpha_3, \beta_3, \frac{v_2^2}{4\varkappa_3}\right)}; \tag{4.64}$$

$$N_1 = \frac{T^* \varphi(\alpha_1, \beta_1, 0) - T_w \varphi\left(\alpha_1, \beta_1, \frac{v_1^2}{4\varkappa_1}\right)}{\varphi(\alpha_1, \beta_1, 0) - \varphi\left(\alpha_1, \beta_1, \frac{v_1^2}{4\varkappa_1}\right)}, \tag{4.65}$$

$$N_2 = \frac{T^* \varphi\left(\alpha_2, \beta_2, \frac{v_2^2}{4\varkappa_2}\right) - T^{**} \varphi\left(\alpha_2, \beta_2, \frac{v_1^2}{4\varkappa_2}\right)}{\varphi\left(\alpha_2, \beta_2, \frac{v_2^2}{4\varkappa_2}\right) - \varphi\left(\alpha_2, \beta_2, \frac{v_1^2}{4\varkappa_2}\right)}, \tag{4.66}$$

$$N_3 = T_0 . \tag{4.67}$$

The velocities $v_1$ and $v_2$ are the solutions of the equations

$$q_1 = \frac{\chi_1}{\varkappa_1} \varphi'\left(\alpha_1, \beta_1, \frac{v_1^2}{4\varkappa_1}\right) \frac{T_w - T^*}{\varphi(\alpha_1, \beta_1, 0) - \varphi\left(\alpha_1, \beta_1, \frac{v_1^2}{4\varkappa_1}\right)} -$$

$$-\frac{\chi_2}{x_2}\varphi'\left(\alpha_2,\beta_2,\frac{v_1^2}{4x_2}\right)\frac{T^* - T^{**}}{\varphi\left(\alpha_2,\beta_2,\frac{v_1^2}{4x_2}\right)-\varphi\left(\alpha_2,\beta_2,\frac{v_2^2}{4x_2}\right)}, \tag{4.68}$$

$$q_2 = \frac{\chi_2}{x_2}\varphi'\left(\alpha_2,\beta_2,\frac{v_2^2}{4x_2}\right)\frac{T^* - T^{**}}{\varphi\left(\alpha_2,\beta_2,\frac{v_1^2}{4x_2}\right)-\varphi\left(\alpha_2,\beta_2,\frac{v_2^2}{4x_2}\right)} -$$

$$-\frac{\chi_3}{x_3}\varphi'\left(\alpha_3,\beta_3,\frac{v_2^2}{4x_3}\right)\frac{T^{**} - T_0}{\varphi\left(\alpha_3,\beta_3,\frac{v_2^2}{4x_3}\right)}. \tag{4.69}$$

In this computation the viscosities have been assumed constant in each zone, and equal to their mean values on the boundaries between the different zones. After the computation of the temperature fields $T_i(\xi_i)$ $(i = 1, 2, 3)$, it is possible to obtain the distribution of the viscosities along the layer, from the formulae

$$\mu_1 = \frac{4x_1}{v_1^2}\int_0^{v_1^2/4x_1}\mu(z)\,dz,$$

$$\mu_2 = \frac{4x_2}{v_2^2}\int_{v_1^2/4x_1}^{v_2^2/4x_2}\mu(z)\,dz,$$

$$\mu_3 = \frac{1}{l}\int_{v_2^2/4x_2}^{l}\mu(z)\,dz,$$

where $l$ is the radius of the third zone, calculated as the length of $r$ when the temperature takes the same value as the initial value (approximately equal).

## 5. Comments and Bibliographical Notes

The papers concerned with different aspects of underground combustion may be divided in three groups: those dealing with models for determining the temperature profile, models for steam flooding and models of combustion.

The models discussed in Sections 1–3 are derived from the general equations obtained in Chapter 4. This presentation follows the paper of Marle (1980), where the reader may find several bibliographical notes.

The examples given in Section 4 are very different. The first is due to Ionescu *et al.* (1985) and Tutunea (1985), and represents a typical determination of a temperature profile. The second example is due to Abatuliev *et al.* (1978), and the third to Kolosovskaia and Rojdestvenski (1972). They may be considered as models for steam flooding. For instance, in the third example $r_1(t)$ and $r_2(t)$ are the boundaries of the combustion front and the condensation front, respectively.

For other aspects of the problem we refer the reader to Gottfried (1965),

Adler (1975), Burger and Sahuquet (1971 a, b, 1972), Legait (1983) and Crookston *et al.* (1979). For practical examples we refer the reader to Burger *et al.* (1984), which is an excellent book for engineers interested in thermal methods of oil recovery.

APPENDIX 1

# Balance Equations for Continuous Media

We consider a material volume $\omega$ of a part $\mathscr{P}$ of a body in its actual configuration. We denote by $\partial\omega$ the boundary of $\omega$ with its unit exterior normal $\mathbf{n}$. Let $\Psi$ be an additive quantity associated with the body. The rate of change of $\Psi$ may depend on both a flux $\mathbf{A}\{\Psi\}$ through $\partial\omega$, and on a production density $Q(\Psi)$ in $\omega$. The general form of the balance equation is then

$$\frac{\mathrm{d}}{\mathrm{d}t}\int_{\omega} \rho\Psi \,\mathrm{d}\omega = \int_{\omega} \rho\, Q(\Psi)\,\mathrm{d}\omega + \int_{\partial\omega} \mathbf{n}\mathbf{A}\{\Psi\}\,\mathrm{d}\sigma\,. \tag{A1}$$

It is well known that, if we consider the motion $x_k = \chi_k(X_m, t)$ for fixed $t$ as a change of coordinates from the reference configuration to the actual one, the relation between $\mathrm{d}\omega$ and $\mathrm{d}\Omega$ ($\Omega$ is the volume of a part $\mathscr{P}$ in the reference configuration) is $\mathrm{d}\omega = J\mathrm{d}\Omega$, where $J$ is the determinant of coordinate transformation.

However, if $\mathscr{P}_n$ is a sequence of parts of the body having only one common point $X$, and $\lim_{n\to\infty} v(\mathscr{P}_n) = 0$ (here $v(\mathscr{P}_n)$ is the volume of $\mathscr{P}_n$ in the actual configuration), it is possible to define the mass density $\rho(\mathbf{x})$ as

$$\rho(\mathbf{x}) = \lim_{n\to\infty}\frac{m(\mathscr{P}_n)}{v(\mathscr{P}_n)}\,,$$

where $m(\mathscr{P}_n)$ is the mass of $\mathscr{P}_n$. From the definition of the mass, it is clear that

$$m(\mathscr{P}) = \int_{\Omega} \rho_0\,\mathrm{d}\Omega = \int_{\omega} \rho\,\mathrm{d}\omega$$

or

$$\int_{\Omega} \rho_0\,\mathrm{d}\Omega = \int_{\Omega} \rho J\,\mathrm{d}\Omega\,.$$

If we suppose that $\rho_0$ and $\rho$ are continuous functions, and bearing in mind that $\mathscr{P}$ is arbitrary, we have in fact the local form of the mass balance:

$$\rho_0 = \rho J \tag{A2}$$

or, equivalently,

$$\dot{\rho} + \rho\,\mathrm{div}\,\mathbf{v} = 0\,. \tag{A2'}$$

Now, using (A 2) we can compute:

$$\frac{d}{dt}\int_{\omega}\rho\Psi\,d\omega=\frac{d}{dt}\int_{\Omega}\rho_0\Psi\,d\Omega=\int_{\Omega}\rho_0\dot{\Psi}\,d\Omega=\int_{\omega}\rho\dot{\Psi}\,d\omega\,.$$

Then, applying the flux-divergence theorem, Equation (A 1) takes the form

$$\int_{\omega}(\rho\dot{\Psi}-\operatorname{div}\ \mathbf{A}\{\Psi\}-\rho\,Q(\Psi))\,d\omega=0\,. \tag{A 3}$$

If the integrand is a continuous function, for any $\mathscr{P}$ at any time $t$, Equation (A3) is equivalent to the differential equation

$$\rho\dot{\Psi}=\operatorname{div}\mathbf{A}\{\Psi\}+\rho Q(\Psi)\,, \tag{A 4}$$

which is the local form of the balance equation.

The momentum and angular momentum balance equations are of the form (see for example Germain (1973) or Truesdell (1969))

$$\frac{d}{dt}\int_{\omega}\rho\,\mathbf{v}\,d\omega=\int_{\omega}\rho\,\mathbf{f}\,d\omega+\int_{\partial\omega}\mathbf{t}\,d\sigma\,,$$

$$\frac{d}{dt}\int_{\omega}\rho\,\mathbf{x}\wedge\mathbf{v}\,d\omega=\int_{\omega}\rho\,\mathbf{x}\wedge\mathbf{f}\,d\omega+\int_{\partial\omega}\mathbf{x}\wedge\mathbf{t}\,d\sigma\,,$$

where **f** is the exterior body force and **t** is the stress vector (this vector is given by $t_i=-t_{ij}n_j$).

With a computation similar to that in (A 4) we obtain

$$\frac{\partial}{\partial t}(\rho\,v_k)+\frac{\partial}{\partial x_l}(\rho\,v_k v_l-t_{kl})-\rho f_k=0\,, \tag{A 5}$$

$$t_{kl}=t_{lk}\,. \tag{A 6}$$

The internal energy balance may be written as

$$\frac{d}{dt}\int_{\omega}(\varepsilon+\frac{1}{2}v^2)\rho\,d\omega=\int_{\omega}(\mathbf{f}\mathbf{v}+r)\rho\,d\omega+\int_{\partial\omega}(\mathbf{t}\cdot\mathbf{v}+q)\,d\sigma\,,$$

where $\varepsilon$ is the internal energy density per unit mass, $r$ is the external radiation per unit mass and $q$ is the heat flux (if we introduce the heat flux vector $\mathbf{q}$, we have $q=-\mathbf{q}\cdot\mathbf{n}$). As previously, the local form of the internal energy balance is

$$\frac{\partial}{\partial t}(\rho\,(\varepsilon+\frac{1}{2}v^2))+\frac{\partial}{\partial x_j}(\rho\,(\varepsilon+\frac{1}{2}v^2)\,v_j-t_{ij}v_i+q_j)-\rho\,(r+f_iv_i)=0\,. \tag{A 7}$$

For a homogeneous mixture of $n$ constituents, each point is occupied by particles of all constituents. Quantities refering to a constituent are denoted by the index $a = 1, 2, \ldots, n$, and there is no summation over these constituents except when explicitly prescribed by the summation sign (see for example Muller (1985)).

The mass, momentum and energy balance equations of a constituent of the mixture differ from the corresponding equations for a single body only by a production term. Indeed, the mass of a constituent may be produced by a chemical reaction, and momentum and energy may be produced by interaction forces and energy exchange between constituents.

By analogy with Equations (A 2′), (A 5) and (A 7), we write the mass, momentum and energy balance equations for constituent $a$ in the form

$$\frac{\partial \rho_a}{\partial t} + \frac{\partial (\rho_a v_i^a)}{\partial x_i} = m_a, \tag{A 8}$$

$$\frac{\partial}{\partial t}(\rho_a v_i^a) + \frac{\partial}{\partial x_j}(\rho_a v_i^a v_j^a - t_{ij}^a) = p_i^a + \rho_a f_i^a, \tag{A 9}$$

$$\frac{\partial}{\partial t}\left(\rho_a\left(\varepsilon_a + \frac{1}{2}(v^a)^2\right)\right) + \frac{\partial}{\partial x_j}\left(\rho_a\left(\varepsilon_a + \frac{1}{2}(v^a)^2\right) v_j^a - t_{ij}^a v_i^a + q_j^a\right)$$

$$= e_a + \rho_a(r_a + v_i^a f_i^a), \tag{A 10}$$

where $m_a$, $p_i^a$ and $e_a$ are the production densities of mass, momentum and energy. They are constrained by the requirements that

$$\sum_{a=1}^{n} m_a = 0, \qquad \sum_{a=1}^{n} p_i^a = 0, \qquad \sum_{a=1}^{n} e_a = 0, \tag{A 11}$$

which express the conservation of mass, momentum and energy in the mixtures as a whole.

Summation of Equations (A 8)–(A 10) over all $a$ leads to the conservation laws of mass, momentum and energy for the mixture as a whole, if (A 11) is used. The simplest illustration of this fact is provided by the mass balance. We define the density and the velocity of the mixture by

$$\rho = \sum_{a=1}^{n} \rho_a, \qquad \rho v_i = \sum_{a=1}^{n} \rho_a v_i^a,$$

and we obtain

$$\frac{\partial \rho}{\partial t} + \frac{\partial}{\partial x_i}(\rho v_i) = 0. \tag{A 12}$$

Usually a chemical reaction $\alpha$ is defined by its intensity $\psi_\alpha$ per unit volume, and the stoichiometric coefficients $\nu_{a\alpha}$ of the reaction. Then $m_a = \sum_{\alpha} \nu_{a\alpha} \psi_\alpha$, and the first condition in (A 11) is in fact $\sum_{a=1}^{n} \nu_{a\alpha} = 0$.

Another form of the mass balance may be obtained by introducing the mass concentration of component $a$:

$$C_a = \frac{\rho_a}{\rho}, \tag{A 13}$$

and the diffusion mass flux

$$\mathbf{j}^a = \rho_a (\mathbf{v}^a - \mathbf{v}). \tag{A 14}$$

Equation (A 8) then has the form

$$\frac{\partial}{\partial t}(\rho C_a) + \frac{\partial}{\partial x_i}(\rho C^a v_i + j_i^a) - \sum_\alpha \nu_{a\alpha} \psi_\alpha = 0. \tag{A 15}$$

The system of $n$ equations (A 8) is equivalent to the system formed by Equation (A 12) and the $n$ equations (A 15), because the latter are not independent. We have:

$$\sum_{a=1}^{n} C_a = 1, \tag{A 16}$$

$$\sum_{a=1}^{n} \mathbf{j}^a = 0. \tag{A 17}$$

If we define the total stress tensor and the specific body force for the mixture by

$$t_{ij} = \sum_{a=1}^{n} (t_{ij}^a - \rho_a (v_i^a - v_i)(v_j^a - v_j)),$$

$$\rho f_i = \sum_{a=1}^{n} \rho_a f_i^a,$$

then the momentum balance for the mixture assumes the form

$$\frac{\partial(\rho v_i)}{\partial t} + \frac{\partial}{\partial x_j}(\rho v_i v_j - t_{ij}) = \rho f_i, \tag{A 18}$$

as in a single body. Note that the total stress of the mixture consists of the partial stresses $t_{ij}^a$ and of the fluxes of momentum of the constituents of the diffusive motion.

A similar argument leads to an energy balance for the mixture, with the definitions

$$\rho\varepsilon = \sum_{a=1}^{n} \rho_a (\varepsilon_a + \frac{1}{2}(\mathbf{v}^a - \mathbf{v})^2),$$

$$q_i = \sum_{a=1}^{n} [q_i^a + \rho_a (v_i^a - v_i)(\varepsilon_a + \frac{1}{2}(\mathbf{v}^a - \mathbf{v})^2) - t_{ij}^a (v_j^a - v_j)],$$

$$\rho r = \sum_{a=1}^{n} \rho_a (r_a + f_i^a (v_i^a - v_i)),$$

$$\frac{\partial}{\partial t}(\rho(\varepsilon + \frac{1}{2}v^2)) + \frac{\partial}{\partial x_j}(\rho v_j(\varepsilon + \frac{1}{2}v^2) + q_j - t_{ij}v_i) = \rho(r + f_i v_i). \qquad (A\,19)$$

Inspection of the previous definitions shows that the internal energy of the mixture is the sum of the internal energies of the constituents plus the kinetic energies of the constituents and of the diffusive motion. The internal energy flux contains the heat fluxes of the constituents, the convective fluxes of the internal and kinetic energies of the diffusion motion, and the influence of the partial stresses on the diffusive motion. Also, the supply of internal energy contains the internal energy supplies of the constituents and the influences of the body forces on the diffusive motion.

With (A 18), Equation (A 19) takes the form

$$\frac{\partial(\rho\varepsilon)}{\partial t} + \frac{\partial}{\partial x_j}(\rho\varepsilon v_j + q_j) = t_{ij}\frac{\partial v_i}{\partial x_j} + \rho r. \qquad (A\,20)$$

The last balance equation is for the entropy. We suppose that the thermodynamic state of fluid depends on $n+1$ independent variables (for example the pressure, the temperature and $n-1$ concentrations). It is known that there is a relation between the differentials of the specific energy, specific volume, entropy and concentrations:

$$d\varepsilon = -p\,d\left(\frac{1}{\rho}\right) + T\,d\eta + \sum_{a=1}^{n} \mu_a\, dC_a,$$

where $p$ is the pressure, $T$ the temperature, $\eta$ the specific entropy and $\mu_a$ the thermodynamic potentials. We then have the relation between derivatives as

$$\rho T\frac{d\eta}{dt} = \rho\frac{d\varepsilon}{dt} - \frac{p}{\rho}\frac{d\rho}{dt} - \sum_{a=1}^{n} \rho\mu_a\frac{dC_a}{dt}.$$

Using (A 12), (A 15) and (A 20), we have

$$Q_i = q_i - \sum_{a=1}^{n} \mu_a J_i^a, \qquad (A\,21)$$

$$\rho\frac{d\eta}{dt} + \frac{\partial}{\partial x_i}\left(\frac{Q_i}{T}\right) - \frac{\rho r}{T} = \sigma, \qquad (A\,22)$$

$$\sigma = Q_i\frac{\partial}{\partial x_i}\left(\frac{1}{T}\right) + \frac{1}{T}(t_{ij} + p\delta_{ij})\frac{\partial v_i}{\partial x_j} - \frac{1}{T}\sum_{\alpha}\psi_\alpha\left(\sum_{a=1}^{n} \mu_a \nu_{a\alpha}\right) \qquad (A\,23)$$

and, by the second law of thermodynamics, $\sigma \geqslant 0$.

The system of balance equations may be completed by adding constitutive equations. Such equations express the energy flux **Q**, the viscous stress tensor $t_{ij} + p\delta_{ij}$ and the chemical reaction intensities $\psi_\alpha$ as functions of the independent variables, and they may be restricted by the second law of thermodynamics. For details we refer the reader to de Groot and Mazur (1962) and Marle (1964).

APPENDIX 2

# Distributions and Sobolev Spaces

Here we give some classical results but without proof. An extensive theory on distributions and Sobolev spaces may be found in Lions and Magenes (1968), Yosida (1965), Adams (1975) and Nečas (1967).

Let $\Omega$ be an open set of the $n$-dimensional Euclidean space $\mathbb{R}^n$. A point $x \in \mathbb{R}^n$ will be denoted by $x = (x_1, x_2, \ldots, x_n)$, and the norm by $|x| = \left(\sum_{i=1}^{n} x_i^2\right)^{1/2}$.

If $\alpha = (\alpha_1, \alpha_2, \ldots, \alpha_n)$, with $\alpha_i$ positive integers, then we denote by $D^\alpha = D_1^{\alpha_1} \cdots D_n^{\alpha_n}$, $D_j = \dfrac{\partial}{\partial x_j}$, the *derivative operator* of order $|\alpha|$. If $G \subseteq \mathbb{R}^n$ we denote by $\bar{G}$ its closure. If $u$ is a function defined on $G$, the *support* of $u$ is the set supp $u = \overline{\{x \in G, u(x) \neq 0\}}$. The function $u$ has *compact support* in $\Omega$ if supp $u \subset\subset \Omega$ (the notation $\bar{A} \subset\subset B$ means that $\bar{A} \subset B$, and $A$ is compact).

A *functional* on a vector space $B$ is a function with scalar values defined on $B$. The functional $f$ is linear if

$$\langle f, ax + by \rangle = a \langle f, x \rangle + b \langle f, y \rangle, \quad x, y \in B, \quad a, b \in \mathbb{C}.$$

The set of all linear and continuous functionals on (the topological vector space) $B$ is the *dual space* of $B$, denoted by $B'$.

Let $B$ be a vector space. A norm on $B$ is any application $f$ of $B$ in $\mathbb{R}$ with the properties

$$f(x) \geqslant 0, \quad \forall x \in B \text{ with } f(x) = 0 \Leftrightarrow x = 0,$$

$$f(cx) = |c| f(x), \quad \forall x \in B, \forall c \in \mathbb{C},$$

$$f(x + y) \leqslant f(x) + f(y), \quad \forall x, y \in B.$$

A vector space $B$ equipped with a norm is called a *normal space*, and we denote its norm by $\|\cdot\|_B$ or $\|\cdot\|$. A sequence $x_n$ from a normed space $B$ *converge* to $x_0$ if $\lim_{n\to\infty} \| x_n - x_0 \| = 0$ in $\mathbb{R}$.

A subset $S$ of $B$ is *dense* in $B$ if any $x \in B$ is the limit of a sequence of elements from $S$. A sequence $x_n$ from $B$ is a *Cauchy sequence* if $\lim_{m,n} \| x_m - x_n \| = 0$. If every Cauchy sequence from $B$ has a limit in $B$, then $B$ is complete and termed a *Banach space*.

If $B$ is a vector space, a *scalar product* is a application $(\,,\,)$ defined on $B \times B$ with values in $\mathbb{C}$ having the properties

$$(x, y) = \overline{(y, x)}$$
$$(ax + by, z) = a(x, z) + b(y, z),$$
$$(x, x) \geqslant 0 \text{ with } (x, x) = 0 \Leftrightarrow x = 0,$$

for any $x, y, z \in B$ and $a, b \in \mathbb{C}$. Here $\bar{c}$ represents the complex conjugate of $c \in \mathbb{C}$. If $B$ is a vector space with the scalar product $(\,,\,)$, the application $\| x \| = (x, x)^{1/2}$ is a norm on $B$. A Banach space with the norm generated by a scalar product is termed a *Hilbert space.*

Let $B'$ be the dual of a normed space $B$. On $B'$ it is possible to define a norm by

$$\| f \|_{B'} = \sup_{\substack{x \in B \\ x \neq 0}} \langle f, x \rangle / \| x \|_B .$$

The space $B'$ with the topology induced by this norm is a Banach space, termed the dual of $B$. If $B$ is a Hilbert space then it may be identified with its dual $B'$.

RIESZ REPRESENTATION THEOREM. *Let $B$ be a Hilbert space; $f \in B'$ iff there exists a unique $x \in B$ such that for any $y \in B$ we have $\langle f, y \rangle = (y, x)$, and in this case $\| f \|_{B'} = \| x \|_{B}$.*

The value under $f$ of $x \in B$ is denoted by $\langle f, x \rangle$, the angle brackets indicating the duality product of $B'$ and $B$.

A sequence $x_n \in B$ is said to *converge weakly* to $x \in B$ if $\langle f, x_n \rangle \to \langle f, x \rangle$, $\forall f \in B'$, and we denote it by $x_n \rightharpoonup x$, or $x_n \to x$ in $B$ weakly. The *strong convergence* ($x_n \to x$ in $B$ strongly $\Leftrightarrow \lim_{n \to \infty} \| x_n - x \| = 0$) obviously implies weak convergence, but the converse holds only in finite-dimensional spaces.

A subset $A$ of a Banach space $B$ is said to be *compact* if any sequence of $A$ has a sub-sequence convergent in $B$ to an element of $A$. $A$ is said to be *precompact* if $\bar{A}$ is compact.

PROPOSITION A1. *The necessary and sufficient condition for $x_n \in B$ to weakly converge to $x$ is that $\| x_n \|$ is bounded and that $\langle f, x_n \rangle$ converges to $\langle f, x \rangle \; \forall f$ in a dense subset of $B'$.*

PROPOSITION A2. *If $f_n$ converges to $f$ in $B'$ strongly, and $x_n$ converges to $x$ in $B$ weakly, then $\langle f_n, x_n \rangle \to \langle f, x \rangle$.*

A sequence $f_n \in B'$ is said to *converge weak star* to $f \in B'$ iff $\langle f_n, x \rangle \to \langle f, x \rangle$, $\forall x \in B$. If the space $B$ is *reflexive* (i.e., if it 'coincides' with the dual of its dual), then the weak and weak-star topologies coincide.

A Banach space is *separable* if there exists a denumerable set of elements which is dense in $B$.

PROPOSITION A3. *If B is separable, any bounded set of B′ is precompact for the weak-star topology.*

This proposition means that if $f_n \in B'$ and $\| f_n \| \leqslant C$, then there exists a sub-sequence $f_{n_\alpha}$ and an element $f \in B'$ such that $\langle f_{n_\alpha}, x \rangle \to \langle f, x \rangle$, $\forall x \in B$. As a consequence:

PROPOSITION A4. *If B is reflexive and the sequence* $x_n \in B$ *is such that* $\| x_n \| \leqslant \leqslant C$, *there exists a sub-sequence* $x_{n_\alpha}$ *and an element* $x \in B$ *for which* $\langle f, x_{n_\alpha} \rangle \to \langle f, x \rangle$, $\forall f \in B'$.

Let $\Omega$ be an open set of $\mathbb{R}^n$. For any positive integer $m$ we denote by $C^m(\Omega)$ the vector space of all continuous functions $\varphi$ on $\Omega$, with the derivatives $D^\alpha \varphi$ of order $|\alpha| \leqslant m$ also continuous. We denote by $C^0(\Omega) \equiv C(\Omega)$, and by $C^\infty(\Omega) = \bigcap_{m=0}^{\infty} C^m(\Omega)$. The subspaces $C_0(\Omega)$ and $C_0^\infty(\Omega)$ are represented by the functions of $C(\Omega)$ and $C^\infty(\Omega)$ respectively, with compact support in $\Omega$.

Because $\Omega$ is open, the functions from $C^m(\Omega)$ are not necessarily bounded in $\Omega$. If $\varphi \in C(\Omega)$ is bounded and uniformly continuous on $\Omega$, then it is possible to define a unique bounded and continuous extension to $\bar{\Omega}$. Then we define by $C^m(\bar{\Omega})$ the space of functions $\varphi \in C^m(\Omega)$ for which $D^\alpha \varphi$ is bounded and uniformly continous for $0 \leqslant \alpha \leqslant m$. $C^m(\bar{\Omega})$ is a Banach space for the norm

$$\|\varphi\| = \max_{0 \leqslant \alpha \leqslant m} \sup_{x \in \Omega} |D^\alpha \varphi(x)| .$$

If $0 < \lambda < 1$, it is possible to define the subspace $C^{m,\lambda}(\bar{\Omega})$ of functions $\varphi \in C^m(\bar{\Omega})$ for which $D^\alpha \varphi$ satisfies in $\Omega$ a Hölder condition of exponent $\lambda$; i.e., there exists a constant $K$ such that $|D^\alpha \varphi(x) - D^\alpha \varphi(y)| \leqslant K|x-y|^\lambda$, $x, y \in \Omega$. $C^{m,\lambda}(\bar{\Omega})$ is a Banach space for the norm

$$\|\varphi\|_{C^{m,\lambda}} = \|\varphi\|_{C^m} + \max_{0 \leqslant \alpha \leqslant m} \sup_{\substack{x, y \in \Omega \\ x \neq y}} \frac{|D^\alpha \varphi(x) - D^\alpha \varphi(y)|}{|x-y|^\lambda} .$$

Note that if $0 < \nu < \lambda \leqslant 1$, there exist the inclusions $C^{m,\lambda}(\bar{\Omega}) \subset C^{m,\nu}(\bar{\Omega}) \subset C^m(\Omega)$.

STONE-WEIERSTRASS THEOREM. *Let* $\Omega$ *be a bounded domain of* $\mathbb{R}^n$. *A subset A of* $C(\bar{\Omega})$ *is dense in* $C(\bar{\Omega})$ *if the following properties are satisfied*:

(*a*) *if* $\varphi, \psi \in A$ *and* $c \in \mathbb{C}$, *then* $\varphi + \psi \in A$, $\varphi\psi \in A$ *and* $c\varphi \in A$,
(b) *if* $\varphi \in A$, *then* $\bar{\varphi} \in A$,
(c) *if* $x, y \in \bar{\Omega}$, $x \neq y$, *then there exists* $\varphi \in A$ *such that* $\varphi(x) \neq \varphi(y)$,
(d) *if* $x \in \bar{\Omega}$, *then there exists* $\varphi \in A$ *such that* $\varphi(x) \neq 0$.

A sequence $\varphi_n \in C_0^\infty(\Omega)$ converges in the sense of the space $\mathscr{D}(\Omega)$ to a function $\varphi_0 \in C_0^\infty(\Omega)$ ($\varphi_n \to \varphi_0$ in $\mathscr{D}(\Omega)$) if the supports of all the $\varphi_n$ are contained in a unique compact set of $\Omega$ and $\varphi_n$, and all their derivatives tend uniformly to $\varphi_0$ and the corresponding derivatives. Then there exists a topology of the vector space $C_0^\infty(\Omega)$ for which a linear functional $T$ is continuous iff $\varphi_n \to \varphi$ in $\mathscr{D}(\Omega) \Rightarrow (T, \varphi_n) \to (T, \varphi)$ in $\mathbb{C}$. The elements of $\mathscr{D}(\Omega)$ are termed *test functions.* A functional on $\mathscr{D}(\Omega)$ is called a *distribution* on $\Omega$, and the set of distributions is the space $\mathscr{D}'(\Omega)$. It is possible to define the convergence in $\mathscr{D}'$: $T_n \to T$ in $\mathscr{D}'$ iff $(T_n, \varphi) \to (T, \varphi)$, $\forall \varphi \in \mathscr{D}$.

If $f$ is a locally integrable function on $\Omega$, it is possible to define a distribution $T_f$ by

$$(T_f, \varphi) = \int_\Omega f(x)\, \varphi(x)\, \mathrm{d}x,$$

which is indeed linear and continuous on $\mathscr{D}$. If $f_1(x)$ and $f_2(x)$ are equal a.e. (almost everywhere; i.e., the set $\{x;\ f_1(x) \neq f_2(x)\}$ is of zero measure), the associated distributions $T_{f_1}$ and $T_{f_2}$ are the same. In fact, the distribution is not associated with a function, but with an equivalence class formed by the functions which are a.e. equal.

If $T \in \mathscr{D}'$, its derivatives are defined by $(D^\alpha T, \varphi) = (-1)^{|\alpha|}(T, D^\alpha \varphi)$, and hence any distribution has distributional derivatives. Moreover, it follows from the definition that derivation is a continuous operation in $\mathscr{D}'$; i.e, if $T_n \to T$ in $\mathscr{D}'(\Omega)$, then $(D^\alpha T_n, \varphi) \to (D^\alpha T, \varphi)$, $\forall \varphi \in \mathscr{D}$.

Let $L^p(\Omega)$, $1 \leqslant p \leqslant \infty$, denote the usual Banach space of (equivalence class of) Lebesgue measurable functions from $\Omega$ to $\mathbb{R}$ under the norm

$$\| f \|_p = \left( \int_\Omega | f |^p \mathrm{d}x \right)^{1/p} \qquad 1 \leqslant p < +\infty,$$

$$\| f \|_\infty = \operatorname*{ess\,sup}_{x \in \Omega} | f(x) |.$$

$L^\infty(\Omega)$ is the space of functions (class of functions) which are bounded a.e.

If $p = 2$, $L^2(\Omega)$ is a Hilbert space, the norm being associated with the scalar product

$$(f, g) = \int_\Omega f(x)\, \overline{g(x)}\, \mathrm{d}x .$$

If $f \in L^p(\Omega)$, $g \in L^q(\Omega)$ with $p, q > 1$, $p^{-1} + q^{-1} = 1$, then $fg \in L^1(\Omega)$, and we have the Hölder inequality

$$\left| \int_\Omega fg\, \mathrm{d}x \right| \leqslant \| f \|_p \| g \|_q .$$

For $1 < p < \infty$ the dual of $L^p(\Omega)$ is $L^q(\Omega)$ with $p^{-1} + q^{-1} = 1$ The dual of $L^1(\Omega)$ is $L^\infty(\Omega)$.

For $1 \leqslant p \leqslant \infty$ and $m \geqslant 1$, we denote by $W^{m,p}(\Omega)$ (or $W_p^m(\Omega)$) the *Sobolev space* of order $m$ on $\Omega$; i.e., the set of all distributional derivatives $D^\alpha u$ of order

$|\alpha| \leqslant m$ belonging to $L^p(\Omega)$. $W^{m,p}(\Omega)$ is a Banach space under the norm

$$\|u\|_{m,p} = \sum_{|\alpha| \leqslant m} \|D^\alpha u\|_p.$$

In particular, $W^{m,2}(\Omega)$ is a Hilbert space, denoted for simplicity by $H^m(\Omega)$, with the scalar product

$$(u, v)_{H^m} = \sum_{|\alpha| \leqslant m} \int_\Omega D^\alpha u(x) \overline{D^\alpha v(x)} \, dx.$$

In particular, the scalar product in $H^1(\Omega)$ is

$$(u, v)_{H^1} = (u, v)_2 + \sum_{i=1}^{n} \left( \frac{\partial u}{\partial x_i}, \frac{\partial v}{\partial x_i} \right)_2 .$$

The completion of the space $C_0^m(\Omega)$ normed as above is denoted by $W_0^{m,p}(\Omega)$, the space $W_0^{m,2}(\Omega)$ is similarly denoted by $H_0^m(\Omega)$ Finally, $W^{-m,q}(\Omega)$, $1 \leqslant q < \infty$, denotes the set of all distributions $u \in \mathscr{D}'(\Omega)$ which can be represented as $u = \sum_{|\alpha| \leqslant m} D^\alpha f_\alpha$, $f_\alpha \in L^q(\Omega)$.

THEOREM A1. *The dual space of* $W_0^{m,p}(\Omega)$ *coincides with the space* $W^{-m,q}(\Omega)$; $1 \leqslant p \leqslant \infty, p^{-1} + q^{-1} = 1$.

It is possible to define $W^{m,p}$ for real $m$. In the particular case $\Omega = \mathbb{R}^n$, $p = 2$, for every $s \geqslant 0$ we define $H^s(\mathbb{R}^n) = \{u \in L^2(\mathbb{R}^n); (1 + |\xi|^2)^{s/2} \hat{u}(\xi) \in L^2(\mathbb{R}^n)\}$, where $\hat{u}$ denotes the Fourier transform of $u$. For $s$ a positive integer, this definition coincides with the previous one.

Now, we suppose $\Omega$ to be a bounded open subset of $\mathbb{R}^n$ with a sufficiently smooth boundary. More precisely, it will be assumed that the boundary $\Gamma$ is an $(n-1)$-dimensional manifold of class $C^\infty$.

This allows us to define the Sobolev space $H^s(\Gamma)$ for any real $s$. Let $L^2(\Gamma)$ be the space of all square-integrable functions on $\Gamma$ with respect to the measure $d\sigma$. Let $\{\chi_i\}$ be a family of local charts on $\Gamma$, and let $\{\theta_i\}$ be a finite partition of unity subordinate to it. For any $u \in L^2(\Gamma)$, we have $u = \Sigma_i \theta_i u$. Let $u_i = \theta_i u \circ \chi_i$. We say that $u \in H^s(\Gamma)$ if $u_i \in H^s(\mathbb{R}^{n-1})$ for all $i$. $H^s(\Gamma)$ is a Hilbert space with the norm defined by

$$\|u\|^2_{H^s(\Gamma)} = \sum_i \|u_i\|^2_{H^s(\mathbb{R}^{n-1})},$$

and it easily shown that neither $H^s(\Gamma)$ nor its norm depends on the chosen family of local charts and the subordinated partition of unity. If $s < 0$, we set $H^s(\Gamma) = (H^{-s}(\Gamma))'$.

For any $u \in \mathscr{D}(\bar{\Omega})$ we can define the *outward normal derivative* of order $j$ on $\Gamma$, $\frac{\partial^j u}{\partial n^j}$. It turns out that the mapping

$$u \to \left\{ u\big|_\Gamma, \frac{\partial u}{\partial n}, \ldots, \frac{\partial^{k-1} u}{\partial n^{k-1}} \right\}$$

can be extended by continuity to all $u$ in $H^m(\Omega)$. More precisely, we have the following theorem:

THEOREM A2. *The mapping*

$$u \to \left\{ \frac{\partial^j u}{\partial n^j}; j = 0, 1, \ldots, \mu \right\}$$

*from* $\mathscr{D}(\Omega)$ *to* $(\mathscr{D}(\Gamma))^{\mu+1}$ *extends to a linear continuous operator*

$$u \to \left\{ \frac{\partial^j u}{\partial n^j}; j = 0, 1, \ldots, \mu \right\}$$

*from* $H^m(\Omega)$ *onto* $\prod_{j=0}^{m-1} H^{m-j-1/2}(\Gamma)$, *where* $\mu$ *is the largest integer such that* $\mu < m - \frac{1}{2}$.

In particular, Theorem A 2 shows that for each $u \in H^m(\Omega)$, $\frac{\partial^j u}{\partial n^j}$ is well defined and belongs to $H^{m-j-1/2}(\Gamma)$. One can also deduce that the space $H_0^m(\Omega)$ coincides with the kernel of the mapping

$$u \to \left\{ \frac{\partial^j u}{\partial n^j}; j = 0, 1, \ldots, \mu \right\},$$

that is,

$$H_0^m(\Omega) = \left\{ u \in H^m(\Omega); \frac{\partial^j u}{\partial n^j} = 0, j = 0, 1, \ldots, m-1 \right\}.$$

EMBEDDING THEOREM. Let $\Omega$ *be a bounded open set of* $\mathbb{R}^n$ *of boundary* $\Gamma$ *which is a manifold of dimension* $(n-1)$ *and of class* $C^\infty$, *and let* $\Omega$ *be on one side of* $\Gamma$. *Then* $W^{m,p}(\Omega) \subset L^q(\Omega)$ *if* $q^{-1} \geqslant p^{-1} - (m/n)$ *and* $W^{m,p}(\Omega) \subset C^k(\bar{\Omega})$ *if* $0 < k < m - (n/p)$, *and the embedding* $W^{m,p}(\Omega) \subset L^q(\Omega)$ *is completely continuous if* $q^{-1} > p^{-1} - (m/n)$.

# References


Abasov, A.A. (1972) 'A Numerical Method for Solving the Problem of Temperature Distribution in an Oil Layer with Steam Flooding' (in Russian), Trudî I.K. A.N. Azerbadjan, Baku.

Abatuliev, F.B., Baklushin, M.B., Erbenkov, Ia.S., and Umarov, U.U. (1978) *Effective Approximation Methods for Solving the Problems of Filtration Theory* (in Russian), 'FAN' Edition, Tashkent.

Abraham, R., and Robbin, J. (1967) *Transversal Mappings and Flows*, Benjamin, New York.

Adams, R.A. (1975) *Sobolev spaces*, Academic Press, New York.

Adler, G. (1975) 'A Linear Model and Related Very Stable Numerical Model for Thermal Secondary Oil Recovery', *J. Can. Petrol. Technol.* **14** (3), 56.

Agmon, S. (1965) *Lectures on Elliptic Boundary Value Problems*, Van Nostrand, New York.

Arquis, E., and Caltagirone, J.-P. (1984) 'Sur les Conditions Hydrodynamiques au Voisinage d'une Interface Milieu Fluide–Milieu Poreux: Application a la Convection Naturelle', *C.R. Hebd. Scéances Acad. Sci. Paris Ser.II* **299**, 1.

Batchelor, G.K. (1970) *An Introduction to Fluid Dynamics*, Cambridge University Press.

Bear, J. (1972) *Dynamics of Fluids in Porous Media*, Elsevier, New York.

Beavers, G.S., and Sparrow, E.M. (1971) 'Compressible Gas Flow Through a Porous Material', *Int. J. Heat Mass Transfer* **14**, 1855–1859.

Bejan, A. (1978) 'Natural Convection in an Infinite Porous Medium with a Concentrated Heat Source', *J. Fluid Mech.* **89**, 97–107.

Bensoussan, A., Lions, J.L., and Papanicolaou, G. (1978) *Asymptotic Analysis for Periodic Structures*, North-Holland, Amsterdam.

Bories, S. (1970) 'Sur les Mécanismes Fundamentaux de la Convection Naturelle en Milieu Poreux', Thèse, Université Paul Sabatier, Toulouse.

Burger, J. and Sahuquet, B. (1971a) 'Combustion a Contre-Courant. Interpretation d'Essais par Modèle Numérique Unidirectionnel' in *Comptes Rendus du Quatrième Colloque ARTEP*, Editions Technip, Paris, pp. 297–320.

Burger, J., and Sahuquet, B. (1971b) 'Chemical Aspects of *in-situ* Combustion. Heat of Combustion and Kinetics', SPE Preprint 3599.

Burger, J., and Sahuquet, B. (1972) 'Laboratory Research on Wet Combustion', SPE Preprint 4144.

Burger, J., Sourieau, P., and Combarnous, M. (1984) *Récupération assistée du pétrole. Les méthodes thermique*, Éditions Technip, Paris.

Burns, P.J., and Tien, C.L. (1979) 'Natural Convection in Porous Media Bounded by Concentric Spheres and Horizontal Cylinders', *Int. J. Heat Mass Transfer* **22**, 929.

Caltagirone, J.-P., (1976) 'Thermoconvective Instabilities in a Porous Medium Bounded by Two Concentric Horizontal Cylinders', *J. Fluid Mech.* **76** (2), 337–362.

Caltagirone, J.-P. (1980) 'Stability of a Saturated Porous Layer Subject to a Sudden Rise in Surface Temperature: Comparison Between the Linear and Energy Methods', *Q. J. Mech. Appl. Math.* **33** (1), 47–58.

Caltagirone, J.-P., Combarnous, M., and Mojtabi, A. (1980) 'Natural Convection Between Two Concentric Spheres: Transition Toward a Multicellular Flow', *Numer. Heat Transfer* **3**, 107–114.

Cattabriga, L. (1961) 'Su un Problema al Contorno Relativo al Sistema di Equazioni di Stokes', *Rend. Sem. Mat. Univ. Padova* **31**, 308–340.

Cheng, P. (1978) 'Heat Transfer in Geothermal Systems', *Adv. Heat Transfer* **14**, 1–105.

Cheng, P., and Minkowycz, W.J. (1977) 'Free Convection About a Vertical Flat Plate Embedded in a Porous Medium with Application to Heat Transfer from a Dike', *J. Geophys. Res.* **82** (14), 2040–2044.

Cheng, P., and Pop, I. (1984) 'Transient Free Convection About a Vertical Flat Plate Embedded in a Porous Medium', *Int. J. Eng. Sci.* **22** (3), 253–264.

Chhuon, B., and Caltagirone, J.-P. (1979) 'Stability of a Horizontal Porous Layer with Timewise Periodic Boundary Conditions', *J. Heat Transfer* **101**, 244–248.

Combarnous, M. (1970) 'Convection Naturelle et Convection Mixte dans une Couche Poreuse Horizontale', *Rev. Gén. Therm.* **108**, 1355–1375.

Combarnous, M.A., and Bories, S.A. (1975) 'Hydrothermal Convection in Saturated Porous Media', in *Advances in Hydroscience*, Vol. 10, Academic Press, New York.

Crookston, R.B., Culham, W.E., and Chen, W.H. (1979) 'Numerical Simulation Model for Thermal Recovery Processes', *SPE J.*, February, p. 37.

de Groot, S.R. and Mazur, P. (1962) *Non-equilibrium Thermodynamics*, North-Holland, Amsterdam.

Elder, J.W. (1967) 'Steady Free Convection in a Porous Medium Heated from Below', *J. Fluid Mech.* **27**, 29.

Elder, J.W. (1968) 'The Unstable Thermal Interference', *J. Fluid Mech.* **32**, 69.

Ene, H.I. (1978) 'Asupra Fenomenelor de Suprafaţă in Cazul Mişcării Fluidelor Vîscoase în Prezenţa unor Medii Poroase' (On Boundary Phenomena in the Case of the Flow of a Viscous Fluid in the Presence of a Porous Medium) (in Romanian), *Stud. Cercet. Mat.* **30** (6), 599–620.

Ene, H.I. (1981) 'On the Thermodynamic Theory of Mixtures', *Int. J. Eng. Sci.* **19**, 905–914.

Ene, H.I. (1984) 'On Thermal Equation for the Flow of a Viscous Compressible Fluid in Porous Media', Preprint INCREST 15.

Ene, H.I., and Gogonea, S. (1973) *Probleme in Teoria Filtraţiei* (Topics in Filtration Theory) (in Romanian), Editura Academiei RSR, Bucharest.

Ene, H.I., and Sanchez-Palencia, E. (1973) 'Equations et Conditions aux Limites pour un Modèle de Milieu Poreux', *C.R. Hebd. Séances Acad. Sci. Paris Ser. A* **277**, 257–259.

Ene, H.I., and Sanchez-Palencia, E. (1975) 'Equations et Phénomènes de Surface Pour l'Écoulement dans un Modèle de Milieux Poreux', *J. Méc.* **14** (1), 73–108.

Ene, H.I., and Sanchez-Palencia, E. (1981a) 'Some Thermal Problems in Flow Through a Periodic Model of Porous Media', *Int. J. Eng. Sci.* **19**, 117–127.

Ene, H.I., and Sanchez-Palencia, E. (1981b) 'Sur la Propagation de la Chaleur dans les Milieux Poreux', *C.R. Hebd. Séances Acad. Sci. Paris Ser. II* **292**, 1981–1984.

Ene, H.I., and Sanchez-Palencia, E. (1981c) 'On Heat Transfer in Porous Media', in A. Verruijt and F.B.J. Barends (ed.), *Flow and Transport in Porous Media*, Proceedings Euromech Delft, p. 143.

Ene, H.I., and Sanchez-Palencia, E. (1982) 'On Thermal Equation for Flow in Porous Media', *Int. J. Eng. Sci.* **20** (5), 623–630.

Fauveau, J., Caltagirone, J.-P., and Combarnous, M.A. (1978) 'Apparition d'Écoulements Thermoconvectifs Multicellularises dans une Couche Poreuse Sphérique', *C.R. Hebd. Séances Acad. Sci. Paris Ser. B.* **287**, 285.

Foiaş, C. (1972) 'Statistical Studies of Navier–Stokes Equations I', *Rend. Sem. Mat. Univ. Padova* **48**, 219.

Foiaş, C. (1973) 'Statistical Study of Navier–Stokes Equations II', *Rend. Sem. Mat. Univ. Padova.* **49**, 9.

Foiaş, C., and Temam, R. (1977) 'Structure of the Set of Stationary Solutions of the Navier–Stokes Equations', *Commun. Pure Appl. Math.* **30**, 149.

Germain, P. (1973) *Mécanique des Milieux Continus*, Masson, Paris.

Gheorghiţă, St. I. (1961) 'The Marginal Stability in Porous Inhomogeneous Media', *Proc. Camb. Phil. Soc.* **57** (4), 84.

Gheorghiţă, St. I. (1966) *Probleme Matematice in Hidrogazodinamica Subterană* (Mathematical problems in Underground Hydrodynamics) (in Romanian), Editura Academiei RSR, Bucharest.

Gheorghiţă, St. I. (1969) 'On the Marginal Stability in Porous Media', *J. Math. Phys. Sci.* **3** (4), 386.

Gor'kov, L.P. (1958) 'Stationary Convection in a Plane Liquid Layer near the Critical Heat Transfer Point', *Soviet Phys. JETP* **6**, 311.

Gossez, J.P. (1966) 'Remarques sur les Opérateurs Monotones', *Bull. Cl. Sc. Acad. R. Belg.* **9**, 1073–1077.

Gottfried, B.S. (1965) 'A Mathematical Model of Thermal Oil Recovery in Linear Systems', *SPE J.*, September, p. 196.

Green, D.W. (1962) 'Heat Transfer with a Flowing Fluid Through Porous Media', Ph.D. Thesis, University of Oklahoma.

Horton, C.W., and Rogers, F.I. Jr (1945) 'Convection Currents in a Porous Medium', *J. Appl. Phys.* **16**, 367–370.

Ingham, D.B., Merkin, J.H., and Pop, I. (1982) 'Flow Past a Suddenly Cooled Vertical Flat Surface in a Saturated Porous Medium', *Int. J. Heat Mass Transfer* **25** (12), 1916–1919.

Ingham, D.B., Merkin, J.H., and Pop, I. (1983) 'The Collision of Free-Convection Boundary Layers on a Horizontal Cylinder Embedded in a Porous Medium', *Q. J. Mech. Appl. Math.* **36**, 313–335.

Ionescu, V., Tutunea, I., and Turta, A. (1985) 'Influenţa Frecării Intermoleculare a Particulelor Fluide în Mişcarea unui Fluid Vîscos într-un Mediu Porous I, II, III' (The influence of the intermolecular friction on the flow of a viscous fluid through a porous medium) (in Romanian), *Stud. Cercet. Meca. Aplic.* **44** (3)–(5).

Kinderlehrer, D., and Stampacchia, G. (1980) *An Introduction to Variational Inequalities and Their Applications*, Academic Press, New York.

Kolosovskaia, A.K., and Rojdestvenski, V.A. (1972) 'Temperature Field in a Porous Elastic Medium with a Reaction Zone', in *Filtration in the Elastic Case and Thermodynamics* (in Russian) Vol. 94, NEDRA, Moscow.

Kvernvold, O., and Tyvand, P.A. (1979) 'Nonlinear Thermal Convection in Anisotropic Porous Media', *J. Fluid Mech.* **90** (4), 609–624.

Ladyzhenskaya, O.A. (1969) 'The Mathematical Theory of Viscous Incompressible Flow', Gordon & Breach, New York, English translation, Second Edition.

Landau, L., and Lifshitz, E., (1971) *Mècanique des Fluides*, Mir, Moscow.

Lapwood, E.R. (1948) 'Convection of a Fluid in a Porous Medium', *Proc. Camb. Phil. Soc.* **44**, 508–521.

Lebedev, V.V., and Sprint, E.A. (1972) 'Filtration and Temperature in Multiphase Flow', in *Filtration in the Elastic Case and Thermodynamics* (in Russian), Vol. 94, NEDRA, Moscow.

Legait, B. (1983) 'Interpretation de Certains Types d'Écoulements Diphasiques en Milieu Poreux a Partir des Écoulements dans des Capillaires', Thèse, Université de Bordeaux I.

Levy, Th., and Sanchez-Palencia, E. (1975) 'On Boundary Conditions for Fluid Flow in Porous Media', *Int. J. Eng. Sci.* **13**, 923–940.

Lindstedt, A. (1883) *Mem. Acad. Imp. Sci. St Petersburg* **31** (4).

Lions, J.L. (1969) *Quelques Méthodes de Résolution des Problèmes aux Limites non Linéaires*, Dunod, Gauthier-Villars, Paris.

Lions, J.L. (1978) 'On Some Problems Connected with Navier–Stokes Equations', in *Proc. Conf. on Nonlinear Evolution Equations*, Madison, Mont., pp.51–84.

Lions, J.L., and Magenes, E. (1968) *Problèmes aux Limites Non Homogènes et Applications*, Vol. I, Dunod, Paris.

Malkus, W.V.R., and Veronis, G. (1958) 'Finite Amplitude Cellular Convection', *J. Fluid Mech.* **4** (3), 225–260.

Marle, C.M. (1964) 'Introduction a l'Etude Thérmodynamique des Systèmes hors d'Equilibre', Thèse, Université de Paris.

Marle, C.M. (1980) 'Simulation Numérique de la Combustion *in situ*', Institut Français du Pétrole, Ref. 27918.

Marle, C.M. (1982) 'On Macroscopic Equations Governing Multiphase Flow with Diffusion and Chemical Reactions in Porous Media', *Int. J. Eng. Sci.* **20** (5), 643–662.

Matheron, G. (1967) *Eléments pour une Théorie des Milieux Poreux*, Masson, Paris.

McKibbin, R., and O'Sullivan, M.J. (1980) 'Onset of Convection in a Layered Porous Medium Heated from Below', *J. Fluid Mech.* **96** (2), 375–393.

McKibbin, R., and O'Sullivan, M.J. (1981) 'Heat Transfer in a Layered Porous Medium Heated from Below', *J. Fluid Mech.* **111**, 141–173.

McKibbin, R., and Tyvand, P. (1982) 'Anisotropic Modelling of Thermal Convection in Multilayered Porous Media', *J. Fluid Mech.* **118**, 315–339.

McNabb, A. (1966) 'On Convection in a Porous Medium', in *Proc. 2nd Australian Conf. on Hydraulics and Fluid Mechanics,* Auckland, New Zealand, p. C 161.

Mojtabi, A., and Caltagirone, J.P. (1978) 'Convection entre Deux Cylindres Coaxiaux en Régime Laminaire Permanent', *Int. J. Heat Mass Transfer* **21**, 261–268.

Mojtabi, A., and Caltagirone, J.-P. (1979a) 'Étude de la Stabilité d'un Écoulement de Convection Naturelle dans un Espace Annulaire Horizontal', *J. Méc.* **18** (2), 225–241.

Mojtabi, A., and Caltagirone, J.-P. (1979b) 'Energy Stability of a Natural Convective Flow in a Horizontal Annular Space', *Phys. Fluids* **22** (6), 1208–1209.

Mojtabi, A., and Caltagirone, J.-P. (1980) 'Analyse du Transfer de Chaleur en Convection Mixte Laminaire Entre Deux Cylindres Coaxiaux Horizontaux', *Int. J. Heat Mass Transfer* **23**, 1369–1375.

Muller, J. (1985) *Thermodynamics,* Pitman, London.

Muskat, M. (1946) *The Flow of Homogeneous Fluids Through Porous Media,* Ann Arbor, Mich.

Na, T.Y., and Pop, I. (1983) 'Free Convection Flow past a Vertical Flat Plate Embedded in a Saturated Porous Medium', *Int. J. Eng. Sci.* **21**, 199–210.

Nečas, J. (1966) *Equations aux Dérivées Partielles,* Les Presses de l'Université de Montreal, pp. 103–126.

Nečas, J. (1967) *Les Méthodes Directes en Théorie des Équations Elliptiques,* Edition de l'Académie, Prague.

Nigmatulin, R.I. (1979) 'Spatial Averaging in the Mechanics of Heterogeneous and Dispersed Systems', *Int. J. Multiphase Flow* **5**, 353–385.

Poliševski, D. (1979) 'Steady Natural Convection in Porous Media', INCREST Preprint Series in Mathematics 41.

Poliševski, D., (1984) 'Homogenization of Thermal Flows in Porous Media', INCREST Preprint Series in Mathematics 55.

Poliševski, D. (1985a) 'Weak Continuity in Convection Problems', *J. Math. Anal. Appl.* **110** (1), 51–58.

Poliševski, D. (1985b) 'The Evolution Darcy–Boussinesq System (A Weak Maximum Principle and the Uniqueness)', *Comment. Math. Univ. Carolinae* **26** (1), 181–183.

Poliševski, D. (1985c) 'Steady Convection in Porous Media–I. The Solutions and Their Regularity', *Int. J. Eng. Sci.* **23** (7), 733–739.

Poliševski, D. (1985d) 'Steady Convection in Porous Media–II. The Case of Low Rayleigh Numbers and Asymptotic Expansions', *Int. J. Eng. Sci.* **23** (7), 741–749.

Poliševski, D. (1985e) 'Steady Convection in Porous Media–III. The Structure of the Solautions', *Int. J. Eng. Sci.* **23** (7), 751–757.

Poliševski, D. (1986a) 'On the Existence and Uniqueness of Non-Stationary Convection in Porous Media', *An. St. Univ. Al. I. Cuza*, s. I, **32** (3), 75–84.

Poliševski, D. (1986b) 'On the Homogenization of Fluid Flows through Porous Media', *Rend. Sem. Mat. Univ. Politecn. Torino* **44** (3), 383–393.

Pop, I., and Cheng, P. (1983) 'The Growth of a Thermal Layer in a Porous Medium Adjacent to a Suddenly Heated Semi-Infinite Horizontal Surface', *Int. J. Heat Mass Transfer* **26** (10), 1574–1576.

Quintard, M. (1983) 'Stabilité d'une Zone Instationnaire de Diffusion Thermique dans un Milieu Poreux: Analyse Linéaire, Modélisation Numérique', *J. Méc. Théor. Appliquée* **2** (5), 751–768.

Quintard, M., and Prouvost, L. (1982a) 'Instabilités de Zones de Diffusion Thermique Instationnaires en Milieu Poreux', *Int. J. Heat Mass Transfer* **25** (1), 37–44.

Quintard, M., and Prouvost, L. (1982b) 'Stability of Nonstationary Thermal Diffusion Zones in Homogeneous Porous Media; Verticals Throughflow Effect', in Proc. 7th *Int. Heat Transfer Conf.,* Paper HC39, Vol. 2, pp. 341–346.

Rudraiah, N., and Musuoka, T. (1982) 'Asymptotic Analysis of Natural Convection Through Horizontal Porous Layer', *Int. J. Eng. Sci.* **20** (1), 27–39.

Sanchez-Palencia, E. (1980) *Non-Homogeneous Media and Vibration Theory,* Lecture Notes in Physics Vol. 127, Springer-Verlag, Berlin.

Schlüter, A., Lortz, D., and Busse, F. (1965) 'On the Stability of Steady Finite Amplitude Convection', *J. Fluid Mech.* **23** (1), 129–144.

Schubert, G., and Strauss, J.M. (1982) 'Transitions in Time-Dependent Thermal Convection in Fluid-Saturated Porous Media', *J. Fluid Mech.* **121**, 301–313.

Schwartz, L. (1965) *Méthodes Mathématiques pour les Sciences Physiques,* Hermann, Paris.

Serrin, J. (1959) 'On the Stability of Viscous Fluid Motions', *Archs Ration. Mech. Anal.* **3** (1), 1.

Slattery, J.C. (1967) 'Flow of Viscoelastic Fluids Through Porous Media, *A.I. Ch. E. J.* **13** (6), 1066–1071.

Slattery, J.C. (1969) 'Single-Phase Flow Through Porous Media', *A.I. Ch.E. J.* **15** (6), 866–872.

Smith, K.T. (1970), 'Formulas to Represent Functions and Their Derivatives', *Math. Ann.* **188**, 53–77.

Strauss, J.M. (1974) 'Large Amplitude Convection in Porous Media', *J. Fluid Mech.* **64** (1), 51–63.

Tartar, L. (1980) 'Incompressible Fluid Flow in a Porous Medium–Convergence of the Homogenization Process', Appendix of Sanchez-Palencia (1980).

Tchearni, I.A. (1948) *Underground Hydromechanics* (in Russian), Izd. Nauka, Moscow.

Temam, R. (1977) *Navier-Stokes Equations,* North-Holland, Amsterdam.

Truesdell, C. (1969) *Rational Thermodynamics,* McGraw Hill, New York.

Tutunea, I. (1985) 'Discontinuous Phenomena in Mechanics' (in Romanian), Ph.D. Thesis, University of Craiova.

Whitaker, S. (1967) 'Diffusion and Dispersion in Porous Media', *A.I. Ch. E. J.* **13** (3), 420–427.

Whitaker, S. (1969) 'Fluid Motion in Porous Media', *Ind. Eng. Chem.* **61** (12), 15–28.

Williams, W.O. (1978) 'Constitutive Equations for Flow of an Incompressible Viscous Fluid Through a Porous Medium', *Q. Appl. Math.,* October, 255–267.

Wooding, R.A. (1963) 'Convection in a Saturated Porous Medium at Large Rayleigh Number or Péclet Number', *J. Fluid Mech.* **15** (4), 527.

Yamamoto, K. (1974) 'Natural Convection About a Located Sphere in a Porous Medium', *J. Phys. Soc. Japan* **37** (4), 1164–1166.

Yosida, K. (1965) *Functional Analysis,* Grundlehren Vol. 123, Springer-Verlag, Berlin.

# Index of Symbols

The number opposite each symbol is the page on which it is defined or explained

# Subject Index

9 789027 722256

GPSR Compliance

*The European Union's (EU) General Product Safety Regulation (GPSR) is a set of rules that requires consumer products to be safe and our obligations to ensure this.*

*If you have any concerns about our products, you can contact us on ProductSafety@springernature.com*

In case Publisher is established outside the EU, the EU authorized representative is:

Springer Nature Customer Service Center GmbH
Europaplatz 3
69115 Heidelberg, Germany

**Batch number: 10607214**

Printed by Printforce, the Netherlands